Environmental Factors in Transportation Planning

Environmental Factors in Transportation Planning

Paul Weiner
University of Connecticut

Edward J. Deak
Fairfield University

Lexington Books
D.C.Heath and Company
Lexington, Massachusetts
Toronto London

Published simultaneously in Canada.

Printed in the United States of America.

International Standard Book Number: 0-669-83055-0.

Library of Congress Catalog Card Number: 73-39342.

To Robin and Alice and Karen,
P.J., and Bryan
In hopes of a better environment
for them and all.

Contents

List of Tables

List of Figures

Foreword

One of the more emotional policy issues confronting local governments over the last decade has been that of the desirability and location of new transport facilities, particularly major highways cutting through or near major urban centers. In my own experience perhaps this emotion was most strongly borne in on me by one particular incident: at an international conference I was once accused by a high-ranking state transportation official of advocating nothing less than "anarchy and undermining of the principles of democratic government" for suggesting that more public hearings might be desirable during the early planning phases associated with major new highway construction.

More tangible evidence of the depth and complexity of these feelings is the extent to which potentially important segments of major urban highways systems are now unable to be built because of the objections, expressed both legally and politically, of various community groups. San Francisco, as is well known, pioneered in this. But it has hardly been so limited: New Orleans, Boston, Seattle, New York, and Philadelphia are just a few of the other cities where similar problems have arisen.

The questions involved are, of course, not easy. They involve very difficult trade-offs. New highways *do* bestow benefits as well as costs and the benefits are enjoyed not only by those who use highways (e.g., new highways often improve the environment in an adjacent neighborhood by reducing the amount of through vehicular traffic). And, of course, the costs associated with new highways are often borne by those who don't use highways to any great extent as well as by those who do. The full incidence of all these costs, moreover, is often difficult to ascertain, though we are finally beginning to understand some of the complexities.

It is to furthering our understanding of these issues, both conceptually and empirically, that Professors Weiner and Deak have devoted this book. Specifically, they attempt to quantify the relevant costs and benefits through questionnaires, applied experimentally to "leadership groups" concerned with such problems in the state of Connecticut. They hope and believe, moreover, that the basic technique has wider applicability and might even be adaptable to determining the attitudes of an entire population through random sampling.

Among their many contributions, Weiner and Deak provide us with some additional insights into important differences in group attitudes. We learn, for example, that rural residential groups, as might be expected, seem more concerned about open space and land-use planning than urban groups and, conversely, that urban dwellers seem more concerned about jobs, income and community effects. Similarly, we discover that highway engineers seem more aware of the potential beneficial effects of new highways than city planners while the latter group is far more aware of the potential aesthetic and community difficulties associated with highway construction. There are, of course, few surprises in these findings; nevertheless, it is reassuring, both about the techniques applied and the validity of the conventional wisdom, to possess somewhat more documented evidence on these points.

Not all problems of environmental planning in transportation are, of course, solved by Weiner and Deak—and they would certainly be the last to claim so. They make a valiant effort to cope with the extremely difficult, perhaps even insolvable, problem of defining a suitable metric for combining the many diverse characteristics and concerns that enter into an evaluation of the total effects of new highway construction. But they will not satisfy all tastes. For example, to some their emphasis will seem too much preoccupied with what one might call "middle-class" environmental concerns and insufficiently aware of lower-income problems of poverty and job creation. Others will, I suspect, record just opposite reactions.

Similarly, better understanding of the environmental trade-offs will be beneficial only to the extent that we can modify governmental and financial institutions so as to better implement on environmental goals—or at least to alleviate the more adverse consequences of new facility development. The essence of the environmental problem in transportation planning is recognition of the fact that in the past we have *not* taken into account all the costs, social and indirect, of new transport developments. In particular, as highway users we have not been paying the full bill. A natural place to turn for some of the additional funding needed to alleviate the adverse environmental aspects of highway development would be, of course, highway trust funds, both federal and state, that are now so commonly suggested for diversion to financing other activities.

But it would be too much to expect that Professors Weiner and Deak could solve all of these problems in one short study. It is enough that they have contributed greatly to furthering our understanding of how to approach and better measure these concerns. And, of course, it is only as our understanding of these problems increases, as it slowly but surely has been and will, that we as a society will identify the means and institutional innovation to better cope and solve these problems in actuality.

John R. Meyer
President, National Bureau of
Economic Research

Preface

Within recent years the general public has become more aware of the sophisticated nature of the interaction between transportation programs and their socio-environmental consequences. This concern has developed as the result of both the disturbing growth of physical evidence indicating the extent of environmental deterioration and the warnings of impending disaster issued by the scientific community. In this crisis atmosphere it is important that additional research efforts be devoted toward the revelation of the functional relationships which exist in this area that are crucial to the continuation of socially optimal decision-making. In recognition of this need, our study is designed to form one link in what the authors hope will eventually be a long chain of empirically based investigations.

In the narrowest sense the research effort was directed towards developing an integrated method of analyzing the many effects of various forms of transportation systems. We also intended to convey a sense of urgency and indicate the degree of difficulty attached to the types of choices which must be made. If it were necessary to distill the essence of the study into a single statement, it would be that in selecting an appropriate transport alternative the policy-maker is also accepting a certain mix or trade-off of social gains and losses. If this or any project is to be undertaken *(and the latter question may ultimately be the most important in the long run)* then the basic concern is the sacrifices which society must make in all forms in order to enjoy the use of the facility.

Any list of thanks to people or groups who provided assistance in this project must be headed by a word of praise for the Connecticut Department of Transportation, which agreed to fund the initial study on which this book is based. Under the able direction of the liaison officer, Mr. Harley Davidson, the Department was extremely cooperative in providing us with both the time and information we required. Also essential to our work was the individual effort and cooperation supplied by each of the directors of Connecticut's thirteen Regional Planning Agencies. It was through this professionally dedicated group of men and women that we were able to secure the active participation of the sample respondents. Special notice must be directed toward the research assistance provided by Mrs. Anne Fisher. Her efforts contributed to the substance of the literature survey in several areas. Finally, only those who have worked on a project of this type can appreciate the indispensable technical service which is provided by the executive secretary. We consider ourselves fortunate for having Mrs. Rosalind Fuchs in this capacity. She performed her task with a degree of patience and dedication which far exceeded the amount normally expected.

As we look back on the results of our study we can only hope that the information which we have developed will be of assistance to others who are striving for environmental quality.

Paul Weiner
Centerville, Cape Cod, Mass.

Edward J. Deak
Sagamore Beach, Cape Cod, Mass.

Introduction

The increase in public concern regarding the "quality of life" has intensified the need for the rational identification, measurement, and evaluation of environmental "impacts" which appear as the result of transportation planning. This book is an attempt to provide an analytical framework and system of measurement for taking these factors into account. It outlines a set of techniques to facilitate investment decision making and provides an actual demonstration of the data collection and interpretation process. After developing an evaluative model, the text describes the field research which was designed to test empirically a number of hypotheses relating to:

1. The adequacy of having citizens, through direct participation, place "weights" on the transport generated externalities which lack an organized market and hence have no price
2. The determination through "sensitivity analysis" of regional variations in attitudes, values, or concerns, and
3. The investigation of a number of alternative techniques which would enable planners to account for environmental factors systematically within the planning process

The model emphasizes the evaluation of environmental "impacts" as contrasted with other approaches that establish a quality dimension by estimating the degree to which goals or objectives have been achieved. The extreme difficulty in reaching a concensus on "goals and objectives" is the primary explanation for choosing this "impact" oriented approach. The "impacts" are formally dealt with on the basis of trade-offs of values and attitudes rather than absolute concerns. To develop this relative concept, the model is designed to put the citizen in the position of establishing a set of priorities for both the beneficial and detrimental impacts of transportation investment.

The book is divided in two major parts. The first focuses upon the role of environmental factors in transportation planning with particular attention being paid to the highway corridor location process. The magnitude and frequency of highway expenditures as contrasted with other modes of transportation requires this special attention. The second part of the book provides a case study illustrating the use of the methodology and one approach to its application through field research. Given the engineering costs of transportation investment and the current techniques regarding the estimation of direct user benefits, the primary question to be resolved is: How can environmental factors which are not reflected in the market be integrated with the calculation of dollar impacts?

No new approach to the problem of nonmarketable externalities can be developed without an adequate understanding of the state-of-the-art. Chapter 1 provides a detailed survey of the literature including cost-benefit analysis, economic impact studies, studies centering on intangibles, formal transportation models, and systems analysis. The principal conclusion of this review is that current techniques fail to consider adequately the total implications of transportation planning. Within each evaluative process, the weakness manifests itself in one or more of three possible ways:

1. The approach fails to consider all impacts
2. The approach fails to solicit public cooperation and participation in evaluating the effects relative to each other in a manner that can be aggregated
3. The approach is unrealistically subjective, thereby obscuring the internal workings of the decision-making process from public review, and consequently hindering a post-decision evaluation of the technique

This deficiency served as a starting point for the development of a set of structural and operational hypotheses around which a methodology evolved that was aimed at testing their validity. The goals which guided the development of the model were twofold: first, to identify and measure the social, economic, and environmental factors that are relevant to the decision-making process in transportation, and second, to construct a systematic framework within which these factors could be evaluated *relative* to each other. Once this framework is established, it would then be possible to demonstrate how the information may be interpreted in order to yield an optimal predictive capability. The reader should be aware, however, that although this information is quantified, no single number should be relied upon solely in making a planning decision. Possibilities of error must be recognized when dealing with a single decisional number. On the other hand, any departure from the direction in which quantification leads a planner should be carefully evaluated and explained.

Chapter 2 provides an introduction to the methodology by detailing the generalized attributes which are common to all review techniques. Finally, the theretical framework relies heavily upon a matrix or balance sheet type of presentation.

Briefly, the technique utilized to evaluate the trade-offs is to create an artificial market for environmental factors through a two phase review process by first assessing the physical impact in terms of dollar compensation and amortizing it over time if necessary, and second, the actual evaluation by a set of relative weights as determined by community participation. Although the compensation need not be paid, the weighted magnitude must enter in the flow of decision making as if it were to be paid so as to reflect the total impact of the proposed corridor. The "impacts" are organized from the prospective of the individual, groups of individuals, such as conservationists or historical societies and the community or some combination of the three. In the event that these calculations prove too expensive in time and money to be included within the planning process, alternative uses of the methodology are discussed.

Part 2 of the book demonstrates the utilization of the techniques designed for empirical investigation. The field research, per se, provided a set of findings that may be of significant use to planners even without reference to the analytical framework. The principal empirically based conclusions of this research are as follows:

1. Considerable variations in transportation related concerns exist among geographically definable regions within the test state
2. Certain factors, especially those that pertain to economics, health, and safety are consistently rated as important in all the regions tested
3. There is some evidence that a difference exists in transportation-related attitudes between individual citizens and local, organized groups
4. Through citizen participation, a relative evaluation of transportation investment can be ascertained by the administration of a questionnaire
5. A format can be developed that will allow for the inclusion of these findings within the decision-making flow
6. Local representatives can be encouraged to participate in a manner that serves a mutual learning experience; i.e., for both the planner and citizen.

Methodology Utilized in Field Work

The instrument used to determine the impact weights was a twenty page questionnaire that was directly administered to the participants.[a] The questionnaire included six categories, with five potential impacts included in each. The six categories were looked at separately from a potential detrimental and beneficial impact. Each impact was described in a general manner that indicated the event would presumably occur. There was also a general category which allowed the respondent to directly compare each of these broad categories. An abstract transportation example was given at the beginning to provide the proper frame of reference.

The source of the thirty categorical impacts came first from an extensive review of previous highway route-location hearings and the concerns that these revealed over a considerable period of time. A second input included national and state attitude studies, while the final source was a close examination of existing academic research. In the questionnaire, respondent's attitudes were indicated by having them make appropriate notational responses. This procedure employed a two stage rating system that was designed to provide a guide to the intensity system.

Three steps were involved in the administration and evaluation of the questionnaire: (1) it was tested by administering it separately to the thirteen regional

[a]Additional information regarding the questionnaire is available from the authors upon request.

planning directors and to the planning staff of the Bureau of Highways of the Connecticut Department of Transportation; (2) it was then administered, in group sessions, to representatives from the thirteen regional planning areas in the state of Connecticut; and finally (3) the tabulated results were returned to each of the planning directors in order that the weights and relative rankings might be reviewed by him. The purpose of the ensuing analysis was to provide descriptive material that could help explain the relative concerns and the actual or potential variations in attitudes within the region.

Levels of Application for Regional Results

This book suggests three potential levels of use for the information gathered by the field work in the area of community impacts. The first use is most general and is simply to provide broad indicators of community concerns to guide transportation planners. These attitudes may be similar to or different from those already held by them. There is some evidence resulting from earlier studies in various states that differences do exist. A second potential use of the information is to develop a community profile. The final approach involves both the quantification of impacts and the construction of actual trade-offs for formal decision making.

In the first instance, where this technique is used in its most general form, those community values that have high priority can and should be revealed. This can best be accomplished by going into the potentially impacted area early in the planning process. Within the study, and as indicated above, it has been determined that certain views held by transportation agencies concerning community priorities may contrast sharply with actual attitudes. For example, a highway planner may feel that a community holds historic sites in very high regard. An actual determination of community attitudes may show just the opposite. Such errors can arise from either a lack of information or the existence of inaccurate information that is supplied by influential community groups that do not reflect the local value system over time. This is not meant to imply that minority views and the potential disproportionate impact of transportation investments on a minority of the community should be ignored. Such an initial procedure is designed only to determine in a general manner a hierarchy of impacts for particular communities to be used in the early phases of the decision-making flow where the route alternatives are devised. It is simply a technique to provide the decision maker, who will still make subjective judgments, with more information.

The second level of use for the information outlining the relative importance of the nonuser effects of transportation investment is the community profile technique, where a descriptive balance sheet of community impacts is developed. In this approach, the impacts on individuals, groups, and the community are arranged in general credit and debit categories. This step is designed to enlarge on the first level by assisting the existing route-location process. This is accomplished through the introduction of further information in an organized manner. The basic problem, however, of actually developing trade-offs is also avoided in this second level, leaving the bulk of the evaluative work to subjective decision making.

In the third level of use which is developed in this text, the information gathered from community residents can be utilized in a numerical framework. It needs to be emphasized again that it is recognized that there is no single number on which an investment decision can or should be based; yet a technique that enables the decision maker to estimate numerical trade-offs has to be developed in order to minimize the subjective elements in decision making. This is a bold approach because many are convinced that certain environmental effects cannot be quantified primarily because a market does not exist for them. As indicated earlier, the approach utilized in this book is to create such a market (actually, a quasi-market) by costing the physical act needed to account for the impact and weighting this by the importance of the compensation to the community. For example, if a proposed highway pollutes a lake, the physical effort required to correct or avoid this pollution is determined. Finally, this dollar figure is weighted by the community's evaluation of the relative importance of this correction factor. If the lake is deemed essential to the community by its citizens in terms of water supply, beauty, or any other reason, it will receive a high weight and, consequently, become a strong factor in the corridor-location selection. If, on the other hand, the lake has an extremely low importance to the community (it may already be polluted), the physical compensation figure would be weighted by a lower relative figure. The resulting index number is entered into the decision-making flow through a matrix format where the item is evaluated relative to other impacts.

Review of Operational Hypotheses

It would appear from the analysis of the tabulated results that some preliminary comments can be offered regarding the substantiation of the operational hypotheses. In particular, several of the hypotheses could be considered as verified in full or in part. An initial hypothesis concerned the physical ability of the participants to perform the operation and develop a set of relative impact weights. That this operation can be performed by representatives of the public is evidenced by the 200 valid responses that were obtained at various regional planning meetings. For the most part, the people were not confused by the mechanical operation and were willing to cooperate with a project that they felt would lead to substantial benefit for their particular region. The point is recognized that the participants can be considered to be above average in their ability to perform the operations. However, the hypothesis regarding the public's ability to participate in such an activity seems in part established. The remaining task would be to simplify the techniques, so that the questionnaire could be administered on a random sample basis, if this particular method is chosen as the future selection procedure to follow.

The second hypothesis offered was that participants were sufficiently aware of their own community attitudes to articulate these in a way that would facilitate transportation investment utilizing highway route location as an example. Again, it appears from the analysis of the results that this hypothesis has been partly established. This is especially true of those impacts that were fairly stable in their rating. If some set of values were not the underlying factors that governed the pattern of selection, the responses would demonstrate considerably less consistency and be more randomly located than they have been observed to be.

The third hypothesis stated that significant variations in the weights assigned to the particular items were to be expected among the participating regions. This particular hypothesis is again substantiated by the extent of place variation. This is most fully revealed in chapter 6. The extent of variation is often accounted for by recognizable area characteristics as described by the three sample regions presented in chapter 5. To make the point clear, an urban region, a semiurban region, and a rural region were abstracted for illustrative purposes. Consequently, the evidence of variation is significant in that it lends support toward the plausibility of the hypothesis.

The fourth hypothesis states that the respondents possess the capability to feel differently regarding the potential beneficial as opposed to the potential detrimental effect as it might appear with respect to the same item. Evidence that this hypothesis has also been established is also given in chapters 5 and 6. In no instance was the rank order identical for the beneficial and detrimental effects and, in many cases, there were substantial variations in rank assigned to a particular item for the beneficial as opposed to the detrimental effect. Consequently, the importance of the item varies substantially within a region from the rank assigned on one side as opposed to the other in the weighting process.

The fifth and final hypothesis concerned the ability of the planners to review the results and to offer comments regarding their validity within the appropriate region. Substantiation of this hypothesis can only be hinted at through the quality of the testimony offered by the planning director as it relates to the analytical profile of his particular region. However, it should be noted that the plausibility of the four previous hypotheses offers the planner the opportunity for analysis based on regional characteristics, because the extent of variations itself was significant enough to facilitate this type of review.

Experience and Conclusions

On the positive side, it was found that the regional planners were most eager to cooperate and participate in this program. These planners may represent a possible avenue for future research of this type. In addition, the administration of the questionnaire itself tends to make community representatives aware of the problems involved in the investment decision-making procedure. It was found that, although this particular questionnaire was difficult, with proper administration approximately 90 percent of the respondents were able to perform the operations. Moreover, by interviewing groups of people, considerable time savings could be realized with only minor reductions in efficiency.

Regarding the limitations of the results, there was some question as to whether there should have been a specific rather than a general highway example used as the frame of reference. This was the most persistent criticism of the test instrument offered by the participants because at times they found it somewhat difficult to view highway construction without reference to a specific corridor. A general approach was followed because a specific example would have tended to exclude too many possible combinations of concerns. A second limitation is that the repetition of this procedure may be extremely costly in a state larger than Connecticut. Third, all of the impacts in each category were assumed to occur. This raised a question of probability versus possibility of occurrence in the minds of the respondents. This may, at times, have led to confusion. Fourth, there is always the problem of whether the respondent's answers change with time and whether he is ever fully aware of his own value system. Fifth, problems of the representativeness of the sample group arise through the implication that the planners and members of the planning commission actually reflect the attitudes of the average citizen. Some comments on this point were provided by the planning directors during the follow-up phase. Sixth, the design questions were not separated from the corridor-location questions. Such a separation might have provided a more optimal questionnaire format. This should be done if the approach is to be repeated in the future. Finally, it may be further argued that the list of thirty impacts excluded some items that may be more important to members of the community or that the grouping of impacts may have affected the final output.

It is apparent at this stage that further research is needed in the area of establishing the magnitude of community impacts. Investigators must move from a general review to the evaluation of specific routes. A new questionnaire may need to be developed that could be administered effectively and easily by highway personnel and responded to by the general public. What is contemplated is something like a "social" origin and destination survey.

Another avenue or research concerns how and at what stage to work this amended procedure into the decision-making process. Because the decision-making effort is a flow with each succeeding step constrained by the preceding decision, some of these nonuser factors would have to be introduced at every stage of official evaluation. Finally, because the responses could change with time depending on surrounding elements, a method would have to be developed that would allow for a continual updating of the general results.

Part I: Survey of the Literature, the Nature of the Problem, and Methodology

1 Survey of the Literature

1.1 Cost-Benefit Analysis

The earliest attempt at accounting for the total impact of highway construction owes its origin to the welfare branch of economic theory. This area of research is directed toward the evaluation of changes in the existing economic system by viewing their desirability with respect to an established set of socially accepted goals. This marriage of ethics (or what should be) with fact (what really is) forms the heart of welfare analysis. In making these evaluations the goal that has been dominant for application in the field of highway analysis is the maximization of economic well-being for society as a whole.

Practical evaluation of welfare maximization requires a unit of measurement by which alternative conditions can be meaningfully compared. While the actual personal level of satisfaction itself should ideally serve as the evaluation criterion, this is not possible since the unit of measurement, its meaning between persons, and the actual calculation process are open to question. As a substitute measure, economists have generally assumed that the level of real per capita income and changes in it yield a good first approximation of social impacts. Although faced with some severe limitations, the common practice has been to apply the "dollar test" in determining the worth of a project. Generally, starting from a given set of initial conditions,[a] any alteration in the pattern of resource use must be so structured that the final level of social satisfaction resulting from the reorganization is greater than the original level.

When stated in these broad terms, this rule of resource allocation (i.e., project evaluation) appears quite simple. However, transference into the area of dollar estimates of welfare and doubts concerning implicit assumptions tend to complicate the statement. The general statement, when expressed in dollar terms, implies that each individual has an equal capacity for satisfaction. Such a capacity assumption cannot really be made since it is impossible, given the present state of the art, to make interpersonal comparisons of satisfaction. If a situation occurs where different individuals are injured as well as benefited by a specific resource reallocation, then questions arise regarding the true social effect of the activity. Without some means of making judgments regarding the relative impacts on those positively affected, as opposed to those who are harmed, the only activities that could legitimately be accepted would be those which hurt no one.

[a]Such conditions include assumptions regarding the desirability of the existing social situation.

In recognition of this problem a compensation principle was introduced into welfare economics.[1] Although troubled by its own limitations, the principle was directed at those activities where the beneficiaries used their gain to compensate the injured parties. If, after the compensation took place, there remained some net gain to the beneficiaries, then the activity was deemed socially desirable from a welfare standpoint. This compensation principle has been carried over into the area of cost-benefit analysis and provides part of the background for this form of project evaluation.[2] Specifically, if it can be shown that social well-being, as measured by per-capita income, rose, due to the reallocation of resources, then it is felt that the activity is acceptable from a community point of view. A determination is made by the actual comparison of the known and projected dollar costs and benefits arising due to the project. The net increase in income itself implies the ability to make the compensation.

The use of the dollar metric in the above manner still involves some assumptions which have been open to question by economic theorists: (1) it ignores effects on the distribution of income indicating to whom the benefits accrue and (2) it implies that all costs and benefits can be measured accurately in the marketplace. It is the criticism of the second assumption and the effect it has on highway planning that forms one of the key elements limiting the effectiveness of cost-benefit analysis as a decision-making guide.

Incorporating the information outlined above, Prest and Turvey have provided a working definition of cost-benefit analysis: "Cost-Benefit Analysis is a practical way of assessing the desirability of projects, where it is important to take a long view (in the sense of looking at repercussions in the future, as well as the nearer view) and a wide view (in the sense of allowing for side-effects of many kinds on many persons, industries, regions, etc.) i.e., it implies the enumeration and evaluation of all relevant costs and benefits."[3] This concept of cost-benefit analysis found its first American application in the field of water resources through the Flood Control Act of 1936. Here the existence of net benefits was introduced as the economic justification for government construction of a particular project. The actual accounting procedure was outlined by the act, and its definition of the range of impacts to be considered remains today as the idealized standard to be followed. In particular, investigators were directed to include all costs and benefits "to whomsoever they may accrue." The use of the dollar standard in the area of water resources was relatively straightforward since many of the effects were measured directly in the market, or possessed a market counterpart from which imputations could be made.

The use of cost-benefit analysis in highway construction can be traced to the landmark Oregon study undertaken in the late 1930s.[4] Here an individual state attempted to determine the overall impact of a highway project based on a comparison of the discounted present value of the facility's measurable costs and benefits. At the time the results of such a comparison were considered to provide an extremely scientific and useful tool to aid in the investment decision-making process.[5]

The apparent success of projects of this type and the need for a comprehensive, nationally acceptable standard for evaluating net highway impact, led the American Association of State Highway Officials (hereafter referred to as AASHO) to publish the now famous "Red Book" in 1952 with a revised version issued in 1960.[6] The effect of this publication was to shift the focus of official attention away from analyzing the economic impact of highways from the community or social point of view toward the relative evaluation of highway user benefits compared to capital and maintenance costs. In the introduction AASHO listed the factors they felt should be considered in an overall study of highway economics:

1. Solvency of a system or group of systems of highways
2. Land and community benefits from highways and their improvements
3. Costs of construction or improvement of highways
4. Costs of maintenance and operation of highways and their appurtenances
5. Direct benefits to road users in the form of reduced vehicle operating costs and savings in time on improved highways
6. Benefits to road users in the form of increased comfort and convenience
7. Benefits to road users in the form of overall accident reduction[7]

This outline is notable for its lack of any direct reference to the measuring of social costs of highways and their inclusion in the calculus of investment decision making. Factors, such as air and noise pollution, aesthetics, and the community consequences of human and industrial dislocation, are all omitted from consideration on the dollar cost side. In part, this was a reflection of the highway body's concern during the 1940s and 1950s with the "total highway," which neglected the interaction of the highway with its environment.[8] By concentrating on highway impacts from the point of view of the user, the possibility of seriously miscalculating the economic justification of a project was introduced, where significant community consequences were present. In effect, the basic tenant of cost-benefit analysis, that benefits and costs should be counted "to whomsoever they may accrue," was being denied.

AASHO compounded the problem by considering only items 3, 4, 5 and 6 in the body of the publication. In explaining the narrow scope of their treatment of cost-benefit analysis they noted that "to a limited extent the other three are included in the discussion, but because of their overall character and the general absence of specific values of accepted accuracy it is not possible to enumerate usable values despite their importance."[9] As a result of considering the impact of a highway project from the point of view of the highway user alone, "the method of analysis herein is not an economic analysis in a broad sense and cannot be used as such."[10] However, AASHO did feel that although the method outlined could not be used as a means to "determine the worth of a proposed improvement,"[11] it could be valuable in comparing alternatives when used in combination with other analytical tools. Unfortunately, this method of evaluation, usually known as engineering economy, has in several instances been improperly applied as economic evidence of the net beneficial character of a specific project. Furthermore, in many cases the analysis has generated methodological errors even in this narrow user form.[1 2]

In applying cost-benefit analysis in its broadest form (i.e., from a community point of view) the classification of economic and social factors shown in table 1-1 would prove most helpful:

Table 1-1
Classification of Economic and Social Factors

Benefits	Costs
1. User (of the new highway)	1. Construction
A. Savings to existing traffic	2. Maintenance
(1) Market	3. Operating and administrative
(2) Non-market	
B. Value of the facility to generated traffic	
(1) Market	
(2) Non-market	
2. Nonuser (of the new highway)	
A. User of alternate facilities	
(1) Market	
(2) Non-market	
B. Community effects	
(1) Market } ←	[4. Social Costs]
(2) Non-market }	

The cost categories 1-3 are fairly self-explanatory and the combination of current engineering and economic techniques is capable of providing reasonably accurate estimates of these entries. This particular group of items has in the past served as the source of data for the denominator of benefit-cost ratios (B/C) used in engineering economy studies. The fourth category appears as a result of taking the broad community view of highway impacts. The dollar costs found here comprise estimates of detrimental effects of highways that are not usually accounted for due to their largely nonmarket nature. Factors such as smog, traffic noise, aesthetic blight, and the loss of community park lands or historic buildings are included under this category. These impacts clearly involve a reduction in the level of public welfare, which necessitates their inclusion within the dollar metric framework.

One difficulty in dealing with these factors (other than the measurement problem) involves the way in which such "costs" are to be accounted for. In particular, two alternatives are open to the investigator: either they can be counted as positive factors on the cost side or negative factors on the benefit side. Although both are theoretically acceptable, the final accounting framework can potentially affect the ultimate alternative chosen, depending upon the way in which the benefits and costs are compared.[13] Consequently, some consistent means of handling these social costs must be devised which at least recognizes this possible shortcoming. The procedure adopted in most cases has been to consider social costs as negative benefits and to uniformly apply the accounting procedure to all the relevant alternatives considered. Social costs will henceforth be considered to affect the benefit side in this way.

The benefit side of the ledger is divided into two classifications, which again reflects the community view adopted by the analysis. Benefits are considered according to the groups affected (user or nonuser) and according to whether these factors are valued at their actual dollar rate or whether an imputation must be made to represent the estimated value (market or nonmarket). Total user benefits (market and nonmarket) are estimated by adding the value of the savings in transportation costs associated with an alteration in an existing facility together with the value of the facility to the new traffic generated by it. Positive user savings, to which estimates of market value can be attached, would include: reduced vehicle operating costs, reduced fuel and oil expenditures, and savings in the time of passengers and drivers. Negative benefits in this category would include added vehicle operating costs incurred while traveling more miles at a faster speed over the new route. This general classification of factors has been extensively covered in the AASHO report.

Other user benefits to already existing traffic flows are estimated with greater difficulty. In particular, dollar values associated with reductions in accident costs (or increases) and in the consumer evaluation of the change in comfort and convenience connected with expressway travel require imputations so that a money estimate of the change in community welfare can be gained. The imputed values assigned to these factors are always open to question since, for example, with accidents it involves the estimation of the value attached to human life and the valuation of the output foregone due to the loss of working time. In accounting for comfort and convenience any figure (usually expressed in dollars per mile traveled) will by necessity have to reflect each commuter's personal evaluation of every trip. The time of day, purpose of the trip, and the number of other cars on the road will all have a significant effect on this value.

The estimate of the value of the road to the traffic generated by it follows the same division along market and nonmarket lines. However, less emphasis is usually attached to this aspect since a level of congestion and inconvenience, lower than that experienced on the preexisting facility, must arise before this new demand is induced to enter into the market. One group of studies of a proposed

British expressway system arbitrarily assigned a weight to generated traffic that was one-half the value of items comparable to those associated with existing traffic.[14]

In the second or nonuser category, benefits to users of alternate facilities comprise an element often omitted from consideration in the typical engineering economy study. Users of parallel roads gain considerable benefit from the construction of a new super highway due to the reduction in congestion costs. Furthermore, the fall in the number of accidents experienced on parallel roads caused by the reallocation (i.e., diversion) of traffic patterns can be an important element in the determination of the social impact. Again, some of these estimates can readily be valued in the market, while others require imputations.

The final source of nonuser benefits relates to the impact of the facility on the community as a whole. This group of factors is partly made up of the items previously transferred from the cost side and now recorded as negative benefits. Their tabulation and evaluation within the dollar framework of cost-benefit analysis is considered by many to be the most difficult and poorly executed facet of the accounting framework. Other than the measurement of air noise and water pollution resulting from the construction of a highway, what is required here is a dollar estimate of the social impact of neighborhood and personal disruption caused by the new facility. This includes evaluation of individual costs of those who are forced to move, thereby destroying personally meaningful community "linkages," and the estimation of the aesthetic, health, and safety factors which affect residents of the former community who still remain in the area. In widening the scope of the social welfare index even further, community costs due to effects on conservation, public recreational areas and historic landmarks must somehow be accounted for. Notice must also be taken of the positive social, political, and economic benefits which are also external to the use of the facility. Since many of these nonuser factors do not pass through the market or possess a market counterpart from which a dollar value can be implied, this is obviously no simple task which can be accomplished by the application of accepted procedures. If a valid indicator of the total highway impact is to be derived however, their inclusion is a necessity.

When all the items that are included under the categories above have been taken into account, the result will be a dollar estimate of the yearly flow of costs and benefits of the project for its estimated service life. With this information the investigator must make a comparison of present values of the two flows so that the net (or relative) value of the facility to the community can be determined. To perform the comparison of present and future costs and benefits, a discount factor must be introduced that will reduce the cash flows to a common time dimension. This weighting factor is an essential element of the analysis, since the dollar value of a cost incurred today is greater than the same dollar amount returned as a benefit tomorrow due to the earnings foregone by the employment of the funds in this use as opposed to an alternative one.[b]

In making the actual comparison of costs and benefits, the investigator has four (4) alternative methods available to him.[15]

1. Select (the) project where the present value of benefits exceeds the present value of costs:

$$\frac{b_1}{(1+i)} + \frac{b_2}{(1+i)^2} + \frac{b_s}{(1+i)^2} + \ldots + \frac{b_{n+s}}{(1+i)^n} > \frac{c_1}{(1+i)} + \frac{c_2}{(1+i)^2} + \ldots + \frac{c_n}{(1+i)^n}$$

2. Select (the) project where the rates of the present value of benefits to the present value of costs exceeds unity:

$$\frac{\dfrac{b_1}{(1+i)} + \dfrac{b_2}{(1+i)^2} + \ldots + \dfrac{b_{n+s}}{(1+i)^n}}{\dfrac{c_1}{(1+i)} \quad \dfrac{c_2}{(1+i)^2} \qquad \dfrac{c_n}{(1+i)^n}} > 1$$

3. Select (the) project where the constant annuity with the same present value as benefits exceeds the constant annuity (of the same duration) with the same present value as costs:

$$b > c$$

where

b = constant annuity with the same present value as $b_1 - b_n$
c = constant annuity with the same present value as $c_1 - c_n$

4. Select (the) project where the internal rate of return exceeds the chosen rate of discount:

$$r > i$$

$$\frac{b_1 - c_1}{(1+r)} + \frac{b_2 - c_2}{(1+r)^2} + \ldots + \frac{b_n - c_n}{(1+r)^n} = 0$$

[b]This is an application of the economic concept of opportunity cost. If an investment (in money terms which represents the actual employment of physical resources) is made in project A, then its rate of return must be at least equal to what the resources earned in the next most profitable alternative project (ex., B). If this return is not forthcoming, then it would pay to invest in B rather than A. This alternative rate of return (or discount) factor must be considered in comparing costs and benefits from A at different time periods.

where

c_n = cost per year
b_n = benefit per year
s = salvage value
n = life of project in years
i = chosen rate of interest
r = internal rate of return

Problems with the implementation of the four methods exist. First, methods one through three require the selection of an appropriate interest rate *(i)* with which to discount the estimated benefits and costs, while the fourth requires an interest rate *(i)* to compare to the calculated internal rate. In selecting an appropriate interest rate *(i)* each of four methods has been advocated and/or applied in different studies.[16]

1. The long-term rate on government securities–This has the advantage of being an actual known market interest rate. It has a disadvantage in that it may not truly reflect the community opportunity cost of resources employed.
2. Social time preference–This is an estimate of the community value of present as opposed to future consumption at the margin. It has the dual disadvantage of being both nebulous in its calculation and once calculated, it often yields a figure substantially below the private market interest rate *(i)* which tends to divert funds from private to public use.
3. Social opportunity cost rate–This is an estimate of the rate of return to society as a whole from the use of the resources in their next best alternative. Its disadvantage lies in the fact that it is hard to determine the value of the marginal project displaced by the use of the funds; and even if it could be estimated, both the existing composition of the investment mix and its trend over time must be considered.
4. Social opportunity cost combined with the social time preference–This is a relative comparison using a social time preference to discount benefits estimated at their opportunity cost rate rather than money cost rate. Problems with this alternative are a combination of those associated with the calculation of a social opportunity cost and the estimation of the social time preference.

Some have advocated the use of the fourth method of comparing benefits and costs as a way of reducing the distorting effect of the interest rate *(i)* selected.[17] Here the comparison between internal and chosen rates *(r* vs. *i)* is a necessary one and, by engaging in a sensitivity analysis, an acceptable range can be generated which may actually include several of the estimated values for rates in methods 1-3. However, this method also contains certain assumptions concerning the economy which some economists are unwilling to grant.[18]

A second problem involves the actual selection of the method to compare benefits and costs. As indicated earlier (see note 13), the system of accounting is capable of altering the numerical results if the second method *(b/c)* of comparison is used. Furthermore, the first method *(b > c)*, while it eliminates the difficulty just noted, does present evaluation problems of its own. In particular, the subtraction of *all* costs (whether dollar costs or negative dollar benefits) from the estimated present value of all positive benefits results in a figure which represents the net monetary gain (or loss) to society as a group. The choice is then made from a number of competing alternatives by selecting the one demonstrating the highest net benefits. Serious distortion in the allocation of resources can result under this method if there is a "significant" difference in the "costs" of the proposed alternatives. Unless the net dividend is evaluated relative to the costs, it is conceivable that a project with lesser costs and benefits (which are not, however, proportionally smaller) will be selected. In order to correct for this possibility, it is necessary to compare alternatives in terms of their incremental *b/c* ratio so as to evaluate the marginal gain to society of the alternative containing larger benefits and costs. Only through the application of this alternative criterion can a proper evaluation of the margin of decision be made. However, the problem of accounting for benefits and costs reappears as soon as the ratio method is reintroduced.

A third problem area connected with the use of cost-benefit analysis involves estimated imputations that are made for the value of the travel time saved due to the construction of a new route and the dollar equivalent of the comfort and convenience of the road. The history of past investigations had indicated that these are by far the most important items (in terms of dollar size) considered on the benefit side. As a result, the dollar rate at which these factors are evaluated is a crucial element in both the relative ranking of alternatives and in deciding whether a proposal is even an acceptable one.

Estimates of the hourly rate by which savings in time ought to be measured are complicated due to the fact that not all time savings are valued equally. The value of ten minute travel time saved by a trained surgeon is not the same as the value of the same amount of time for a bricklayer. Moreover, the purpose of the journey will be a key variable in assessing the value of travel time savings. The value of journey-to-work time should carry a different weight than savings in leisure time. Finally, the amount of time saved relative to the total time needed to complete the journey will have an effect on the value of the saving. The savings of ten minutes on a journey of thirty minutes will have a value to the traveler that is different from the time saving of ten minutes on a trip of five hours. Obviously, in aggregating these different categories of savings a serious margin for error can be introduced into the benefit calculation. The AASHO "Red Book" selected a value of $1.55 as an estimate of the dollar value per hour of driving time saved per passenger vehicle.[19] Based on a figure of 1.8 persons per car, this is a value of $0.86 per person (in 1960 prices) per hour. Other estimates have ranged as high as $2.50 for the value of hourly vehicle time savings. On the most heavily traveled routes which carry a sizeable portion of the journey to work traffic of a major metropolis, time savings can be measured in millions of hours per year which provides a significant leeway in the dollar benefit estimates. The sensitivity of the final construction decision with respect to this factor is apparent.

The same type of problem is encountered when a dollar estimate of the extra comfort and convenience per mile of new (or improved) road traveled is added to the benefit side. Again the possible extent of deviation generated by trip purpose, road condition, personal temperament, and so forth, is quite large and results in a wide fluctuation in the net benefit or *b/c* result. These limitations are well known to those who rely upon cost-benefit analysis as an evaluative tool, and as partial compensation, a sensitivity analysis is usually performed to determine the impact on the decision of different dollar values. Even with this compensating factor included, results of the analysis in the transportation field are open to some skepticism from critics who feel that too little is known about the true value of these elements to evaluate them properly.

The fourth problem area is the situation where a proposed project is capable of altering the economic structure of an entire region. Here, the use of an estimate of these cost savings as a measure of project benefits will be inadequate, since most of the gains will be related to the generation of new economic activity. What is called for in this instance is a projection of the economic profile of the region at some future date both "with and without" the transportation change. The net difference between the two estimates serves as a measure of potential benefits.[20] This evaluative technique, which also accounts for changes in the relative price level, is most applicable in areas where the initial level of the economic development is low. Here the greatest chance exists for a major transportation project to bring about this structural change. The principle difficulty of the "with and without" method of determining the economic impact is in making sure that all benefits included for the region are really net benefits to the economy as a whole. Too often, an improvement of facilities in one area will throw the projected industrial distribution between regions out of balance. This can result in the diversion of an existing industry from one area to another, turning what are counted as positive regional benefits into zero national gains.

A fifth limitation of cost-benefit analysis combines a number of criticisms which are aimed at practical deficiencies of the technique. It has already been noted that, although the theoretical background calls for an accounting of all costs and benefits, there is a significant group of economic and social impacts that are not adequately considered. In particular, the nonuser, nonmarket category is often given short shrift, owing to the fact that these involve "intangible" items that possess a "non-quantifiable" nature. Any attempts at imputing a value for the social stress and psychological disorganization incurred by area residents due to the intrusion of a highway facility into a previously cohesive neighborhood would in most instances be nothing more than a guess. Transferring this subjective evaluation into dollar terms further increases the possibilities for error and can result in a severe distortion of the net benefit estimate. Similar arguments have been offered against the inclusion of dollar amounts representing the cost of noise, water, and air pollution. The primary impact of these items is external to the facility and its users. They affect society collectively and, therefore, their importance can only be measured by society acting as a body to express its feeling toward them. Any dollar estimates of these effects is open to question since it does not allow for a market test of acceptability. Finally, some nonuser factors for which accurate dollar estimates can be made are still doubtful since their existence and importance depends upon the subjective evaluation of the impact made by those affected.

To include these items within an evaluative framework requires an expansion of cost-benefit analysis beyond the initial dollar criterion. Some preliminary methods for doing this have been developed and will be discussed in section 1.4 under the heading of "Matrix Forms of Evaluation." In general they all conclude that the analysis is hindered by its dollar dimensional nature which limits adequate consideration of "intangibles." This unidimensional fault reinforces the criticism that the methodology eliminates consideration of the distribution of benefits and costs. Attempts have been made to correct for this latter difficulty by including the distribution of income as an explicit variable in social impact evaluation.[21] However, such a factor itself involves measurement problems and has not gained popular acceptance.

One method which goes beyond the normal dollar metric, and at the same time avoids a matrix presentation, considers a dollar estimate of quantifiables combined with a verbal discussion of intangibles. This attack on the evaluation issue creates some distinct problems of its own. Since the temptation is to accept the extended engineering economy study and use it as the basis for a project recommendation, the usual separation of these conclusions into verbal and numerical sections generally "stacks the deck" against these intangibles. These relative weights never seem to properly consider "non-quantifiables." Actually, the entire burden of blame for this oversight cannot be charged to the investigator. The nature of "intangibles" makes them not only hard to express in dollar terms, but also hinders their measurement in any other standardized form. As a result, it becomes difficult to aggregate these factors into any meaningful index which allows for the proper weighting of intangibles as a group. Consequently, what is actually compared is a weighted dollar measure of all tangibles and an individual unweighted description of all intangibles. Rather than considering them as a cohesive body of factors, the individual impacts usually appear to have less meaning than they might otherwise have. The possible interaction of intangibles to produce an environmental condition that is more harmful and unacceptable to human habitation than the simple linear sum of effects is also not adequately considered. This improper comparison of impacts can result in the construction of a facility which serves suburban highway users while at the same time it destroys the usefulness of the urban environment for permanent living on an acceptable level.

In assessing the value of cost-benefit analysis as an analytical tool, it should be pointed out that it is limited in its usefulness as a technique for evaluating potential impacts. The process itself, due to measurement problems, cannot be used as the final determinant of the investment decision. Rather, it serves as an imprecise tool aimed at listing as completely as possible the potential impact of a facility in dollar terms. The result can be used as an indication but not as proof of the worth of a project to the entire community. Unfortunately, in many circles the approach has received an unwarranted degree of sanctification so that it becomes the ultimate justification (often **ex post**) for a facility. The application of the measure in this way at this time is unwarranted to say the least.

The theoretical foundation of the analysis is sound in that the chosen measure should reflect changes in aggregated social welfare. However, the selection of a dollar index of such alterations may itself prove to be the source of errors. Although money is usually considered as the common denominator of material goods, it may well be that other measurement techniques can be developed which can better indicate the level of social welfare. Indexes of length of life, community participation, tension, distribution (as opposed to level) of income, or some such measure may well provide a clearer indication of potential impacts. Attempts to measure highway impacts along these lines have been made and will be considered in section 1.3.

1.2 Major Economic Impact Studies

One of the basic problems of impact studies is the definition of a region. A useful concept, applicable to estimating highway-related impacts, is that of the economic development region, which includes the development of policies, programs, and actions to move the region from its present economic position toward predetermined economic objectives. Any practical definition of a region must deal with problems of data, and roles played by commuters, residents, firms, and government. Data may be available for political subdivisions, but economic subdivisions (including more than one political unit, but not necessarily coterminous) may be more pertinent for public planning purposes. In delineating a region, the smaller the region and the more integrated it is with the national economy, then the more susceptible it becomes to outside influences that are capable of affecting its economic activity. As a result, there is a problem in establishing a logical framework for comparing data used to measure and project economic change. Basically, there are four levels of data aggregation that are of interest to the regional planner (for both short-run and long-run planning). First, the subcommunity functions as a unit in many respects and has its own distinct problems. The second level of aggregation is the town or city. Third along the scale comes the regional planning area covering several towns. Finally, the planner needs data aggregated to the state level in order to decide how to allocate total funds available within the state, and also for comparisons and coordination with other states and their agencies. There are several alternative ways in which desired planning data can be derived.

Regional Accounts

The decision maker for a region currently faces problems that are relatively more complex than in previous generations. To handle this complexity a great deal of work has been done recently in the area of regional accounting. Such accounts, if properly constructed, can be a very useful tool for decision making. Basically, an account of this type is an empirical framework designed to illustrate the functional relationship between variables. The type of account set up will, of course, depend on the use for which it is designed. However, an intensive regional accounting system can easily be aggregated or factored to suit needs of particular agencies, such as those established for transportation planning.

Regional data have four main uses:

1. They are used to inform local governments as to the effect of regional policies
2. They facilitate comparisons of regional and metropolitan areas
3. They are used for forecasts and projections of the future
4. They describe investment needs and growth patterns for long-range planning

The regional accounting approach does have drawbacks in that it relies heavily on balance of payments data for the region, which may not be available or which may be expensive to collect.

Perloff has done extensive work in the area of regional accounts.[22] Using a regional growth-industrial location framework, he calls for a two-level system of accounts: national accounts covering all regions and accounts for within each region. This would allow greater coordination between differing regions, depending on purposes to be served. To some extent, Perloff follows the economic base theory when he distinguishes between "export" and "derived" industries. The relative regional advantage will determine the location of new productive capacity for "export" industries. Perloff sees the metropolitan area primarily as a labor market. Location of productive activities depends upon high profits and low costs. Therefore, firms are concerned with input access, market access, and transfer costs, which indicate the importance of the transportation network. The most rapid economic growth has occurred in areas where both consumption and production elements are favorable. Perloff wants data showing the year-by-year proportion or "share" of national totals, implying growth in one section at the expense of decline in another. He also discusses the "composition effect" which is the result of a region having a favorable or unfavorable mix of rapid-growth vs. slow-growth industries. Shifts among regions within given industries are due to changes in relative regional advantages for these industries. Thus, regions can grow from either the introduction of faster growing industries or from existing firms obtaining a larger share of the output of a given industry.

Four different approaches can be used in planning, according to Perloff. First, regional accounts and reporting can show "the state of the community," aiding in policy decisions. For this, data are needed on purchases and sales, value added, and income paid out associated with each of the major industry groups within the area (i.e., systematized stock and flow data, including human assets). Secondly, projections of growth in population and economic activities, and projections of structural changes are needed to guide investment decisions, for public services plans, and for private marketing arrangements. Thirdly, impact or interaction analysis is needed to predict policy implications of actions which may be implemented. Finally, evaluation of specific programs, projects, and policy alternatives using interaction or cost-benefit analysis, can aid the planner. Community change can be measured by variations in the volume and stability of economic activities, population, welfare, and productivity. Relative regional advantages determine the location of a new productive capacity. Locational decisions rely on data for manpower characteristics, input-output, income-outgo, and assets or wealth of the region. Also, location depends upon the importance that the industry attaches to the following: transportation, large markets, communications, external economies, labor, and large sites. Thus, it is clear that some form of regional accounts can help the transportation planner assess the impact of a change in the transportation network.

The Economic Base of a Region

Economic base theory has played a major role in regional analysis. Weimer and Hoyt first gave a clear statement of the theory by developing a method for selecting basic industries and determining their relative importance.[23] Starting primarily from consideration of real estate values, they constructed a way of estimating future population by an analysis of the economic base.

The economic base is defined to be the community's export activities that bring in net earnings. Units of measure are always a problem in an economic base study. Employment information may be available, but it overlooks productivity variations and is inapplicable to capital exports. The number of jobs, as well as the payroll amounts, are also important. Another measure is value added, but this entails problems of price adjustment and complex price movements. Furthermore, this method is also deficient in measuring capital exports and commuter activity. Identification is an important aspect of an economic base theory, including the problem of segregation of basic and service elements from "mixed" activities and the definition of economic boundaries for the community. Hoyt's residual method is no longer used since more detailed data is now available and because it assumes a 1:1 ratio between basic and service employment. Obviously the goal is to find this ratio, not to assume it. The size of an area is important, since a small city will have more exports than a large, diversified metropolis. To solve this problem, Andrews suggests that an area be defined as the labor market area of a city or the community range.[24] This would be generally adaptable to all types of communities and bases.

Many studies have used the base ratio concept, which assumes a somewhat constant relationship between the size of the labor force and the size of the population that it supports. Typically the base-to-services ratio is considered to be 1:2 and the base-to-production ratio is about 1:6. Andrews states that the causes of the ratio differing among cities depend on the nature of the economic base itself, geographical location, regional economic cycles, and the economic maturity of the area. There is an implicit assumption that when one of the components is changed, the community will eventually return to original ratios. This implies a concept of the "ratio norm" to use as a bench mark for city planners. Giles and Grigsby point out that empirical estimates might be distorted by the instability of the base ratio,[25] since the growth of a community does not necessitate that the various components of the base increase in the same proportion. However, even with its limitations, the base ratio does permit rough predictions and it is easy to obtain.

Economic base studies are often used as a part of the planning process. Blumenfeld states that the applicability of the basic-nonbasic concept decreases with increasing community size,[26] and that its applicability increases with increasing specialization and division of the labor between communities. Also, the greater the amount of "unearned" income flowing into or out of the local economy, the less useful is the base concept (the base ratio is highest in small, new communities and lowest in large, mature areas). He points out that the identification of export activities for each locality is important for national planning of industrial location, but if local planning agencies use this as a guide to promotion, it will either be ineffective or it will lead to an inefficient national location pattern. Blumenfeld feels that as long as the "non-basic" industries function efficiently, a region will be able to substitute new export industries for those which are declining.

The use of base theory for prediction in linear regression analysis rests on six assumptions:

1. The increase in exports is autonomous
2. The production function for basic activities is unchanged during the adaptation to the rise in exports
3. The marginal propensity to consume and the marginal propensity to import are constant
4. The production function for service activities does not change during the adaptation to increase community demand
5. There is a feasible means by which export activities can be expanded (such as through attraction of new industries resulting from highway improvements)[c]
6. For prediction of population, the job-people ratio is constant.

Any one of these assumptions may easily be violated in reality.

Economic base studies may be useful for transportation planning agencies in order to give a rough assessment of what a community is like and how it may change with transportation improvements, but it is clearly only an indicator and not a precise evaluation.

Multiplier Models

After review of the base-ratio analysis, theoreticians developed a multiplier concept for predictive purposes. Levin[27] computed a "foreign-trade" multiplier *(k)*

$$k = \frac{1}{1 - Y_{end}/Y_t}$$

where

Y_t = total income or value added

Y_{end} = value added in production for the local market

Blumenfeld discusses a multiplier defined to be the ratio of total population to basic employment.[28] Use of the multiplier for population prediction required knowledge of future employment in basic industries, the average living standard, the percentage of goods and services (in value terms) imported, other net money flows into the local economy, the family coefficient of persons in basic and nonbasic employment, and the percentage of people not employed. However, even with this multiplier it is difficult to predict future basic employment, since it is the most variable element of the urban economy.

[c]These five assumptions imply that the only way exports can be raised is through an increase in national demand.

Tiebout expands this, saying that since employment, income, and industrial production vary together, any one of these can be used as an indicator of general economic welfare.[29] Tiebout sets up an income multiplier model where total urban income equals income earned in independent activities plus income earned in dependent activities. He points out that base theory, while it can be thus modified to be consistent with economic theory, still cannot handle the problem of secular growth and is, therefore, essentially a short-run tool. In forecasting, the long-run emphasis is on the export industries. This implies the need for an examination of export industries and feasibility studies to attempt to match industry needs with available resources. However, it is difficult to predict needs of unformed industries.

Hegeland shows that the basic weakness of multiplier theory lies in its assumptions which eliminate real problems involved.[30] This occurs due to the fact that, if the distribution of new and subsequent increments in the income stream are assumed to be given, then the limiting value of the infinite geometrical series is at once known. Hegeland concludes that multiplier theory loses relevance when it is applied to social aggregates, but that it can be fruitfully applied to micro problems, such as an evaluation of the impact of changes within a single stream of an expenditure within a limited sector of the economy. This extension of an economic base theory has limited applicability for transportation planning.

Input-Output Analysis

Regional and intraregional decisions are made by private industry, by local, state, and federal government, and by voluntary service organizations. These decisions may be made to deliberately offset specific activities, or they may be made independently of outside considerations. Hirsch proposes tracing the impact of decisions through a regional intraaction model.[31] This generalized decision-making model uses external forces, the internal private sector and the internal government sector as inputs, with the region's health and well-being as outputs. This well-being can be assessed by considering: (1) per capita personal real income, (2) basic (export) employment stability, (3) net social benefits, (4) economic growth, and (5) amenities of life. However, there must be a model to quantify the impact of alternative policies on these specific dimensions.

With an input-output grid of the general form that Hirsch proposes, it is possible to follow inputs and purchases from purchaser to consumer, showing the dependence of one industry upon another. Isard and Kavesh demonstrate that this method gives a round-by-round description of how an impulse acting on one sector is transmitted to other sectors of same regions and to other regions.[32] Input-output analysis assumes that the initial stimulus (such as an improvement in the transportation network) has final demand and technical coefficients that are known.

Input-output tables are difficult and costly to construct. They include no allowance for substitution of inputs, restraints due to limited resources, or changes in consumption patterns resulting from variations in income.

Since intraregional activities are person and place centered, street sections, land parcels, and persons are appropriate data units for intraregional accounts. However, choice of a unit is complicated by government fragmentation and private industry's location decisions. Generally, it seems best to collect data on the smallest possible unit, and then aggregate according to the desired objectives. Basic data should include land requirements and population characteristics. Land is needed for the provision of public and voluntary services, for private goods and services, for recreation and residential purposes, and for movement. The household account should show locational characteristics of people as both producers and consumers. Government services (such as transportation facilities) need to be related to people and land, or to the demand for such services.

With the desire to design an operational model in the sense that data could be obtained at a reasonable cost, Hansen and Tiebout set up a model for California and its three major subregions.[33] External forces usually lead to greater fluctuations in the level of economic activity than internal forces, especially when the external forces account for large spending (as with defense spending in California). Hansen and Tiebout attempt to determine the impact of shifts in final demand on employment.

Tiebout develops a mathematical conceptual framework of an input-output model and then discusses its limitations.[34] Use of national projection coefficients at the regional level is probably unjustified, as illustrated by the regional differences in fuel demands. The use of an average coefficient is probably not much better, due to vastly differing requirements of industries. The assumption of stable trading patterns for resources overlooks import substitution and varying locational advantages, and the assumption of constant costs implies excess capacity which may not be realistic (at least for some sectors of the region). Tiebout concludes that most operational regional studies have only tentative results stemming from the lack of adequate data and the necessity of making operational assumptions that often may vary from reality.

Transportation-oriented Impact Studies

The choice of the technique to be used in the evaluation of alternative transportation proposals will depend upon the size and potential impact of the project being considered, the time and funds available for the evaluation, and certain concerns of the decision-making body. The total impact of a highway will include changes in land values, reorganization of land uses (creating changes in investment and output), and changes in employment and the labor supply. A new highway may increase opportunities for mobility, fluidity, employment, and recreation, and create a need for more services from public utilities and the government. It is not only the effect of the highway on a community that must be evaluated; we must also consider the community's impact on the transportation system. For example, suburban development depends on highway transportation facilities which permit city workers to live in the surrounding area. Once the highways have been established, shopping centers will grow, creating more pressure for improved transportation facilities. Growing highway needs have led to modifications of the roles of state, local, and federal government with respect to transportation. In the first place, transportation routes and terminal facilities may occupy substantial amounts of urban space, affecting the continuity of neighborhoods. Secondly, capital and operating costs often take a large share of the public budget so that transportation needs must be balanced against other requirements for urban growth. Finally, the transportation system alters the pattern of urban development, indirectly affecting the productivity and "livability" of the region. These factors must be considered along with the efficiency of moving people and goods.

Heenan[35] and Golden[36] have done separate studies to measure changes in land values resulting from an improvement in the transportation network. Their evidence generally supports the thesis that rising land values are associated with increasing accessibility (resulting from improved transportation opportunities). Hinkle and Frye conclude that additional traffic capacity implies that traffic redistributes itself by upgrading, leading to lower accident rates, lower vehicle operating costs, and so forth.[37] They employ an interesting definition of "adverse travel," which is the additional distance that a person will travel in return for a benefit such as time, comfort, or safety.

Stoll has done a very limited study to show that overall average trip length for a region's residents does not seem to change over time.[38] However, he points out that "salient socio-economic factors such as income, occupation, race, household size, auto ownership, and housing density directly affect the length and orientation of trips."[39] This would indicate that changing characteristics of a region would affect the transportation needs.

Dodge has completed the first part of what seems to be a very promising before-and-after study of the impact of a highway on two primarily agricultural counties in northern Wisconsin.[40] Recognizing first that trends continue even without the proposed change, secondly that other factors affect the area, and finally that improvements in one section may be at the expense of another section, he has set up a five-step evaluation scheme: (1) select an area large enough to include control areas with similar economic characteristics; (2) record the level of economic activity in the area *before* and (3) *after* the improvement; (4) take account of extraneous factors causing changes in economic activity regardless of the transportation improvement; and (5) compare traffic volume and patterns serving different types of economic activity. Dodge hypothesizes that the second part of the study will show that the new Interstate route has less trade influence on smaller communities than on larger ones, that the highway improvement will not affect the trade areas of communities which do not lie along the corridor, that the improvement will counteract the tendency toward larger farms (as people work a small farm and have a job in the city), and that the improvement will not affect farm values. Although this study is not directed at an urban area, the results to be presented in part 2 should be very interesting.

Employment trends for the past two decades show rapid growth in the suburbs with a corresponding decline in employment in central cities.[41] For transportation networks, this trend implies a heavy increase in work travel in older suburbs, with a relatively small increase in travel toward the central city. With the exception of New York City, areas with the best transit systems seem to have the most rapid decline in central city employment. Thus, both downtown and suburban traffic will continue to rise, but suburban traffic will increase at a faster rate.

Witheford has undertaken a study concerning highway impacts on the central business district and the suburban shopping center.[42] He attempts to compare the characteristics of the market areas within a fixed travel time from a hypothetical central business district and suburban shopping center. He assumes that an area-wide urban highway improvement takes place, leading to increased travel speeds. As a result, larger market areas can be reached in the same travel time. In a test case for the Niagara Frontier Transportation Study, Witheford found gross family incomes higher in the suburban shopping center area than in the central business district. Retail sales have been increasing in suburbs because of the growth of suburban shopping centers whose success is due to easier parking, pleasant environment, and greater accessibility to consumers.

In a hypothetical case, assume that residential densities decline with increasing distance from the core and that highway speeds increase with distance from the core. "Research has shown that shopping-center trip-generation rated at the residential origin zones are more sensitive to time than distance,"[43] so time is the relevant variable. Higher incomes in suburbs may mean larger purchasing power there, even though the density is lower in the suburbs. If the highway improvement leads to a uniform numerical increase in travel speeds, the increment to the suburban shopping center area is greater than the area added to the central business district. This is shown in figure 1-1. For the market overlap, the proportion of the suburban shopping center now reached from the central business district is less than the proportion of the central business district now reached from the suburban shopping center. The suburban shopping center favoring effect is even stronger if the central business district is located off-center. However, some central business district shopping will not be affected by travel time (office workers, tourists, etc.), and public transit is usually available in central business district areas, offsetting the effect somewhat. The conclusions of this article, although they represent an interesting model, are rather limited in that they only consider the impact on retail sales.

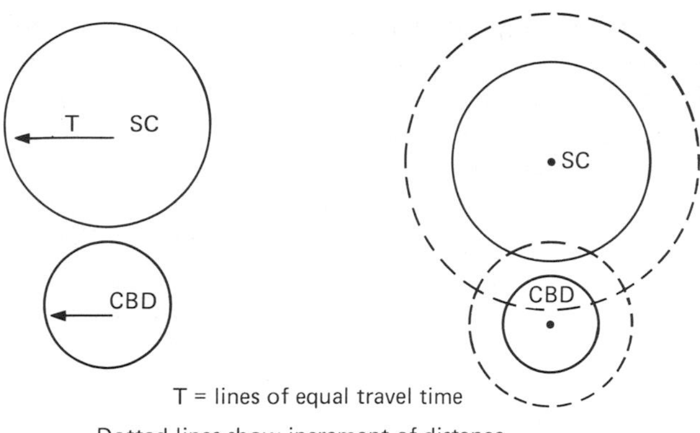

Figure 1-1. Accessibility to Shopping Facilities with Equal Travel Time. Source: David K. Witheford, "Highway Impacts on Downtown and Suburban Shopping," *Highway Research Record*, No. 187 (Highway Research Board, 1967), p. 16.

After surveying several existing works, it is apparent that a considerable amount of time and effort has gone into the problem of evaluating the impact of highway investment projects. It is clear that a transportation system helps to shape the urban area, and that the growth of an area creates pressures which help develop the transportation system. Economic impact studies, whether in the form of regional accounts, economic base studies, multiplier analysis, input-output techniques, or special impact studies, generally consider only the measurable economic factors, without consideration of many of the indirect and intangible effects. As a result, these methods are an imperfect guide to transport planning. The use of any one of the above techniques generates a risk factor that involves the omission of certain community impacts which are relevant to the investment decision-making process. To overlook such impacts raises the possibility of the creation of a transport network which is economically efficient, but socially undesirable.

1.3 Studies Centering on Intangible Variables Which Possess a Nonmarket Dimension

The inadequacy of current evaluative techniques in properly accounting for the intangible impacts of highway construction has led a number of people to devise new types of measurement tools. As pointed out in section 1.1 on cost-benefit analysis, it is difficult in a practical sense to assign dollar estimates to intangible nonmarket factors. Any attempt in this direction is highly subjective and possesses the potential of significantly distorting the resulting value of net benefits. To avoid this problem, it was suggested that separate results be constructed for quantifiable and nonquantifiable factors. Since any verbal presentation of individual nonmeasurable factors would tend to be neglected relative to the more official and authoritative appearing dollar estimate of the impact, this approach was also open to criticism. To partially compensate for this deficiency, several indexes have been developed to present a stronger image of the possible impact of a highway on the community's social structure. In particular, these studies center on the potential disruptive impact of a highway upon those elements of the community structure which are considered most important by the residents. In order to understand the meaning of these indexes, it becomes necessary to view both the source and the process of social interaction.

The concept of sociological disruption and the existence of problems requiring a humanistic tension-reducing approach was examined in detail by Diesing.[44] His conclusions serve as a basic background for the discussion of sociological indexes. In resolving sociological conflicts the primary process is one of identifying and eliminating the source of personal or group stress and strain. Any psychosocial problem (in the present transportation context this would involve adverse community social impact) manifests itself in personal objections to or uneasiness about some element of the present or potential environmental condition. This dissatisfaction arises because of a perceived conflict between the establishment values of the community (or individual) and a force what is altering the preferred status quo. This is not to imply that all change is interpreted as being directed against the community structure. Rather, only those changes that run contrary to the prevailing hierarchy of values and which affect a "significant" number of area residents can be considered as detrimental. The anticipation, estimation of magnitude, and resolution of this stress and strain has been the prime objective of the social indexes.

To estimate the social impact, it becomes necessary to identify the community affected and to list or describe values with which the proposed facility is in conflict. A community value is understood to be any concept or norm that possesses an intrinsic worth for its own sake and which is accepted by a recognizable segment of the area under investigation. This definition would eliminate perverse individual values while including those which, even though they may be unpopular, are held by an established minority of the community. These values can be social, economic, political, historical, geographical, or aesthetic in nature. Each of these value categories then possesses a list of descriptive criteria by which the attainment of the value can be measured.[45] For instance, a proposed transportation facility may have an impact on the structure of social values by altering the use and length of leisure hours, the pattern of neighborhood residences, or the image that persons living in an area have of themselves and their community. If these changes possess a prominent negative aspect, then a given segment of the community can be counted upon to voice their objections to the value conflict they feel is too severe. What the transportation authority must do in order to alleviate these human conflicts is to identify first the source of disruption and then to try and reduce the social stress and strain.

As outlined by Diesing,[46] the resolution of social conflict problems requires an "integrating" technique that attempts to harmonize divergent interests, thereby reducing tensions. The resulting process of mutual adjustment never follows a set formula, since it is a rare instance where a single value is being affected. Rather, the interdependent nature of the value structure usually leads to a multiple disruption of these values. What is required in this case is to try to select a crucial problem area that is the most independent from others and then to reduce the tension there, using the results as a momentum-generating force to continue on to other areas of conflict. The integrating process can be regarded as an attempt at negotiating away the source of stress and strain. In many instances, problems that arise in the area of value conflict are not unsolvable because values themselves are often subject to constant reorganization and fluid change.

In applying this integration approach to the area of highway impacts, several types of value conflicts have been used by different groups as the focal point for tension-reducing activity. One segment of the construction process, which has been considerably altered over time with the recognition of these factors, involves the distribution of information concerning potential routes and regions to be affected. The procedure has changed in most highway departments from one of complete secrecy concerning proposed projects to a comprehensive program of discussion and counseling with community representatives so that accurate, authoritative information and advice can be exchanged between the interested parties. The effect of this interaction has been a reduction in the level of tension caused by fear and rumor. Unfounded accusation concerning the potential impact on both social and economic elements considered vital to the interests of the community are usually eliminated. On the other hand, if open conflicts do exist between local values and regional requirements, then it is just as important to circulate as much information as possible so that these problems can be defined at an early preconstruction stage and the integrating process can be established to reduce the level of community stress and strain.

Another general attempt at identifying potential areas of value conflict involves the use of the "team" concept in the location and design of new highways.[47] In applying this method it is recognized that highways possess several impact dimensions (economic, social, political, etc.) and that the only way to view each potential impact from its proper perspective is to employ a group of investigators, each trained in a different field, to analyze the situation and anticipate the potential impact. Included in the team would be selected representatives from the social sciences, urban planners, architects, design and traffic engineers, together with specialists that might be called for in specific instances, such as conservation experts, navigation authorities, or so on. By working together and in conference with local leaders, many of the tension creating aspects of the facility can be anticipated and attention can be directed to them in early preconstruction stages. The results of such team projects have been quite beneficial, since area residents are usually more than willing to cooperate with highway authorities when they feel that such contacts will lead to the mutually acceptable resolution of recognized conflicts.

While both the distribution of information and the team approach are aimed at identifying and alleviating potential sources of stress and strain, they are still general approaches that focus on the broad spectrum of community values rather than centering upon one or more key aspects which can be used as an indication of the potential intensity of community dissatisfaction. To correct this shortcoming, two indexes have been devised by different authors which demonstrate in quantifiable terms the community impact of a project. One index was constructed by California State highway officials for their use in measuring the possible effect of urban highways.[48] Based upon a workable definition of neighborhood, the index attempted to aggregate specific elements that reflect the area's residential stability and susceptibility to cultural change. By interpreting the value of the index, it was felt that guidelines could be drawn which would aid in both route selection and in the estimation of potential impact.

The assumptions upon which the index is built involve an interpretation of the way community values manifest themselves. It was felt that these values could best be demonstrated and interpreted by the behavioral patterns of area residents. The elements within the community structure which are most important are the ones that people treated as such. In placing these values and behavioral patterns within the proper geographically defined neighborhood, the factors that outline the essence of the neighborhood concept are defined. A neighborhood, as the Hill group views it, could best be evaluated along these cultural and behavioral lines. They felt that a neighborhood was organized so that persons of similar background and values were capable of associating with each other. These working concepts of community values and neighborhood provided the initial hypothesis from which a mobility index was constructed. In particular, "to the degree that population in a neighborhood is stable, the cultural patterns of that neighborhood can be expected to be continuous, persistent and enduring."[49] It was important, then, to find measures of neighborhood stability and receptiveness to change and to aggregate them into index form as a predictor of potential impact.

After due consideration of the problem, the following variables were deemed relevant. As a measure of population stability, the key factors were:

1. The percentage of persons five years old or over who occupied the same residence for five years.
2. The percentage of units twenty years or older which were occupied by the same household during the entire period.

Elements which reflected the susceptibility to change included:

3. Percentage of resident units which are single family.
4. Percent of owner-occupied units.[50]

The mobility index (M.I.) itself is the linear sum of the stability and propensity to change percentages (M.I. = stability and change). A high value for the index would indicate a significant level of cultural continuity that can result in severe tension if disturbed by highway construction. On the other hand, a low value would imply a less stringent objection to the project *along the culturally defined neighborhood dimensions*. The application of the index created data problems necessitating the alteration of the chosen factors and a change in the weighting scheme. The final mobility index was:[51]

$$\text{M.I.} = \frac{\text{percentage of residents}}{\text{same place for 5 years.}} + \frac{\text{ownership and single residence}}{2}$$

In utilizing the information obtained from the index, the highway department was most interested in viewing the effect of freeway location on the character of the neighborhood. However, if behavioral patterns and cultural continuity are important community aspects, then the index can also be used to anticipate areas of potential difficulty. The link with the Diesing articles concerning stress and strain due to value conflict is quite evident.

The major difficulty with the Mobility Index lies in the fact that it is still unidimensional; it centers on only one aspect of a community. As recognized by Hill, the index does not measure the "class" or "status" of the community. It fails to explain the cause of cultural continuity in an area and ignores other values which are important to a highly mobile population. Historic, economic or political values can be injured if the land taken by eminent domain is a key factor in the community structure. Any one of these values can be crucial in creating civic tension and none is adequately represented by the mobility index. Furthermore, either direct discrimination or the continued inability of specific racial, cultural, or religious groups to move from certain residential areas may be the hidden force behind the cultural continuity. Any action which requires a reduction in existing housing barriers may actually be welcomed. In summary, the Mobility Index measures only one factor which should be considered by investigators looking for possible sources of community tension.

An index which attempts to compensate for the problems of the Hill measure was devised by Raymond Ellis.[52] He notes that in evaluating the community impact of highways, most definitions of neighborhood involve both a social and a spatial dimension. In constructing his index, Hill centered upon the social factor of cultural continuity and used it both as a criterion in itself and as an indicator of the spatial factor. On the other hand, Ellis points out that there is a significant interaction between the social characteristic of an area and the actual physical elements of the environment. Any index of community impact based upon a few key variables must take this interaction into account. In particular, a measure of the community effects as opposed to personal consequences of a transportation system must perform two functions: (1) it must spatially and socially define the community and (2) it must accurately gauge the impact of the facility on the area.[53]

Ellis feels that existing neighborhood impact measures fail to fulfill these functions completely. As a result, he offers a substitute index based upon the need for "communication" within the neighborhood. He defines residential linkages as "ties between the housing site of the household and other spatially distinct points which are of importance to the individuals involved."[54] By viewing the number and direction of linkages, a spatial and social definition of the neighborhood can be constructed. Weighting the importance of these revealed lines of communication involves estimating personal satisfaction. However, as Ellis points out, the substitutability of other links and frequency of contact can be used as indicators of their value.[55] Once the direction of communication has been established in an area, highway construction plans can be geared to an avoidance of unwarranted disruptions of the origin and termination points of the interaction. Alterations can take place by changes in land use, the barrier effect of a highway traversing an area, or through changes in the residential structure by bringing into the area new groups who require different linkages.

This index, too, does not provide a detailed picture of the root cause of the linkage, since all changes in communication need not be harmful. However, it can safely be stated that, generally, the greater the linkage disruption, the more people affected; and consequently, the greater the possibility of tension due to value conflict. The failure of the index lies in the fact that a low disruptive effect on communication does not necessarily imply a limited impact by the facility on the area as a whole. Air pollution and traffic noise are external elements of a system which possess harmful neighborhood aspects. The barrier effect of a raised embankment facility may preclude future expansion of linkages into nearby areas and be unsatisfactory as a result. The increased use of neighborhood streets to gain access to expressway approach ramps can create serious health and safety problems. Finally, the aesthetic effect of the highway on the remaining structures can be a source of stress and strain. None of these possibilities is adequately predicted by a residential linkage index.

In summing up the existing state of the art with respect to measuring the community effect of changes in the transportation system, it should be noted that most of the attempts center on a limited number of key factors as indicators of potential difficulty. The items chosen by Hill (cultural continuity and receptiveness to change) and Ellis (residential linkages) are quite ingenious in that they are intimately connected with and reflect other aspects of community life which are considered important by area residents. Their main weakness, however, is that they lack sufficient flexibility to serve as a useful indicator of disruptive potential in a majority of cases. Situations can arise where severe tension is created in an area, without it being reflected by the index. As a partial measure of one or two key factors, the indexes are helpful, but as a reliable guide to action in all instances they fall short. Other ways must be found to evaluate the intangible economic and social effects of highways so that they may be properly weighted in the decision-making process.

Another limitation of both of the approaches, involves the fact that they illuminate only the harmful or tension-creating community impacts. Obviously, not all the external effects of highways are detrimental. In many cases the redevelopment aspects of urban road construction can have significant positive social impacts. However, no provision exists in either approach for noting these beneficial effects and allowing their comparison to the "linkage" or "continuity" variables harmed by construction. Unless these aspects can be included there is no way to evaluate the relative impact of the facility.

1.4 Matrix Forms of Evaluation

Developing future transportation facilities necessitates plan design, testing, evaluation, selection, correction, and implementation. The alternatives to be evaluated should cover plans with a fairly wide range of capital costs, since this would provide a choice of plans offering different levels of service. This implies a trade-off between the investment of government funds and individual user services. Figure 1-2 shows the evaluation range to be considered for total transportation systems.

Gendell gives useful definitions of terms related to evaluating projects.[5 6] *Values* represent a system of preference that governs society's actions. *Goals* are

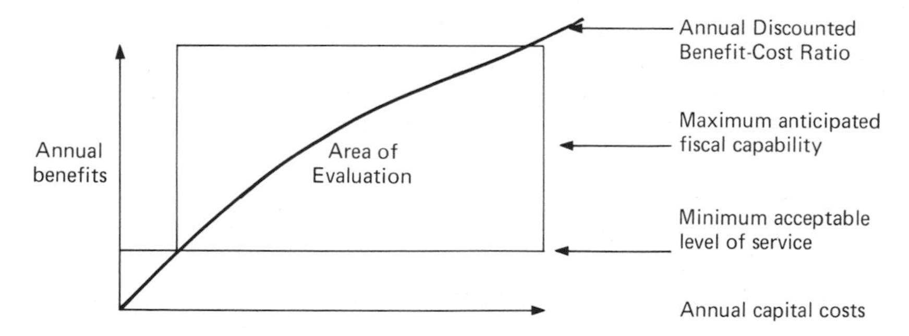

Figure 1-2. Area of Evaluation. Source: David S. Gendell, *Evaluation of Alternative Transportation Systems*, Lecture 92 (Bureau of Public Roads), p. 2.

ends strived for, while *objectives* are components of goals or steps necessary for goal attainment. *Evaluation criteria* are standards to use for comparing the ability of plans to attain goals and objectives. *Policies* may be defined as definitive courses of action toward plan implementation, while *programs* are coordinated segments of plans (subsets of policies).

Determination of community goals and objectives is one of the first essentials for any plan evaluation method. Schimpeler and Grecco propose evaluation based on community structure, by establishing a weighted hierarchy of community decision criteria. They postulate setting up a committee for weighting criteria.[57] The members would be the general community decision makers, composed of "direct and indirect influentials including popular public officials and representatives of commerce and industry who are influential in controlling development decision" and interested parties.[58] Testing their method, they found ten community goals that were further subdivided into objectives. The goals were: (1) public safety, (2) public utilities and transportation, (3) economic development, (4) cultural development, (5) health, (6) education, (7) welfare, (8) recreation, (9) political framework, and (10) housing. The objectives were then weighted by four alternative techniques and a high correlation between the weights obtained was observed. To be useful, the objectives must be stated precisely and related in a measurable way to alternative physical development proposals.

Hill compares quantitative and qualitative objectives, giving a hierarchy of classes of measurement scales to be used in measuring quantities.[59] The nominal scale can be used to classify and number entities; the ordinal scale ranks activities, while an interval scale allots equal intervals between entities and indicates their distance from the (arbitrary) origin. The most complex measure is the ratio scale which has equal intervals between entities and indicates their distance from a nonarbitrary origin. The validity of any measure depends on the extent to which it measures what it is supposed to measure. Thus, the usefulness of a particular scale should be judged on this basis.

In general, evaluative procedures assume that goals and objectives not only have been identified, but also have been weighted. Hill suggests that these weights could be determined iteratively through a process of interaction among elected officials, administrators, planners, and other groups, all reacting to explicitly stated objectives.[60]

The Twin Cities study used the Value Profile Method for evaluating goal satisfaction.[61] This is a graphical comparison, with a chart showing the extent to which various goals are judged to be satisfied by alternative plans. The method does not include any weighting of the goals, consequently it gives no indication of the way to choose alternatives which have crossing profiles as shown in figure 1-3.

The Southeastern Wisconsin Planning Commission used a Rank-Based Expected Value Method for nonquantifiable as well as quantifiable factors.[62] This

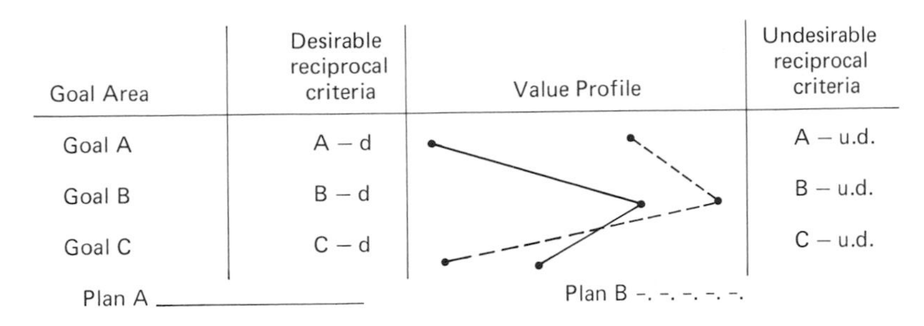

Figure 1-3. An Illustration of the Value Profile Method.

method first ranks plans according to their ability to satisfy the objectives and then ranks the goals in order of importance. The plan score is adjusted by the probability of implementation. The method multiplies the rank value for each objective by the rank value for the plan's ability to meet the objective, then sums across the objective. That score is then multiplied by the probability of implementation, and the plan with the highest value is chosen. This method raises the objection that plans are ranked by the order of importance and little consideration is given to intensity.

Assuming that objectives have been established and weighted, Schimpeler and Grecco propose two decision models for use in plan evaluation.[63] In the effectiveness matrix technique, the set of community planning objectives i represented by G_j, $j = 1, 2, \ldots, n$, where n is the total number of decision criteria considered. The plans under evaluation are p_i, $i = 1, \ldots$, m. For each objective G_j there is a numerical utility value, $u_j (j=1, \ldots ,n)$ which is determined by a weighting process. The weights must be adjusted so that

1. $$\sum_{j=1}^{n} u_j = 1.$$

If e_{ij} = "effectiveness" or the measure of probability that objective j can be achieved if plan i is adopted, and u_i = "plan utility" or the measure of the total utility of plan i based on the evaluation of that plan relative to all n objectives, then the effectiveness matrix E is defined as in figure 1-4.

		Criteria
		$G_1, G_2, \ldots, G_{j'}, \ldots, G_n$
Alternative Plan	$P_1,$	$e_{11}, e_{12}, \ldots, e_{1j'}, \ldots, e_{1n}$
	$P_2,$	$e_{21}, e_{22}, \ldots, e_{2j'}, \ldots, e_n$
	P_i	$e_{i1}, e_{i2}, \ldots, e_{ij}, \ldots, e_{in}$
	P_m	$e_{m1}, e_{m2}, \ldots, e_{mj'}, \ldots, e$

Figure 1–4. Effectiveness Matrix.

For the decision model, additive utilities are used, so that

2.
$$u_i = \sum_{j=1}^{n} e_{ij} u_j \quad (i=1, \ldots, m)$$

where u_i is the total utility for plan i. The highest u_i indicates the plan to be chosen. This model may be shown in vector notation as

3.
$$u = E_u \qquad \text{or}$$

4.
$$\begin{bmatrix} u_1, \\ u_2, \\ \cdot \\ \cdot \\ \cdot, \\ u_i, \\ \cdot \\ \cdot \\ \cdot, \\ u_m \end{bmatrix} = \begin{bmatrix} e_{11}, e_{12}, \ldots, e_{1n} \\ \cdot \\ \cdot \\ \cdot, \\ \\ \\ \\ \\ \\ e_{m1}, e_{m2}, \cdots e_{mn} \end{bmatrix} \begin{bmatrix} u_1 \\ \cdot \\ \cdot \\ \cdot, \\ \\ \\ \\ \\ \\ u_n \end{bmatrix}$$

This effectiveness matrix may be made into a scoring model to consider different socioeconomic groups in the region. To do this, let

5. $$u_i = u_i^1 + u_i^2 + \ldots + u_i^k + \ldots + u_i^p \text{ and}$$

$$u_i^k = a_k \sum_{j=1}^{n} u_j^k e_{ij}^k$$

where u_i = the total score for alternative plan i

u_i^k = the total score for alternative plan i as determined by socio-economic group k, ($k = 1, \ldots, p$)

u_j^k = the criterion weight for objective j as determined by the kth socioeconomic group

e_{ij}^k = the value of plan i relative to criterion j as seen by the kth socioeconomic group, and

a_k = the fraction of the region's population represented by the kth socioeconomic group.

Again, the decision model is $u = E_u$, where the score $(u_i 0)$ for each plan is weighted according to the socioeconomic group composition.

Schimpeler and Grecco point out that this is a suboptimization process, since it is dependent upon the policy-making and financial processes, which need well-defined criteria at a higher level. The method assumes cardinal utility values in the optimum allocation of community resources. However, the model does consider the importance of the community decision structure and the formulation and weighting of goals and objectives.

Jessiman et al., also try to find a way to consider explicitly the significant nonmonetary benefits and costs along with those which have dollar estimates.[64] They consider a radial highway to be built in a rapidly growing suburban corridor. The alternative choices would be available for the location and length of the new highway, design standards, and whether or not to build it at all. They propose a five-step technique:

1. Itemize the objectives for the facility (assuming that the alternatives have been proposed)
2. Define the best measure for each objective
3. Weight the objectives (relatively)
4. For each objective, evaluate the way the alternative meets the objective using a scale such as a utility curve
5. For each alternative, sum the values for all objectives and choose the one with the highest total

Use of a utility curve for weighting rather than a relative scale has advantages in that the scale is not affected by the inclusion of another alternative. Jessiman's technique allows judgment and subjective feeling to be included in an organized framework and provides for the mixing of subjective measures with those derived mathematically. Of course, it faces difficulties in determination of the utility curves.

Gendell summarizes the Schimpeler-Jessiman-type analysis in a value matrix, which he sees as an extension of the rank-based expected value method.[65] Once again assuming that goals, objectives, and evaluative criteria have been established and weighted, data are collected pertaining to the relation of alternative plans to the individual criteria (in money, quantifiable, or qualitative terms). The plans are then ranked on a 0-10 scale according to how well they satisfy the criterion, and the ranks are summed across several criteria for a plan score. The matrix itself uses only goals, objectives, and evaluative criteria to which people can be expected to give realistic value indications. Other factors enter as constraints or trade-offs. The constraints help alleviate the problem of using one value for the weight of each criterion regardless of the performance of various plans (the marginal utility problem). Trade-offs involve cost effectiveness as measured by the plan score. Gendell also gives an eight-step example using a value matrix. Although the value matrix can be a planning aid, there seems to be no completely satisfactory method of plan evaluation. However, even if faulty, some systematic, comprehensive evaluation method should be used in assessing the ability of various transportation proposals to fulfill the regional transportation goals.

Hill defines rational planning as a process for determining future action by utilizing scarce resources in such a way as to maximize the expected attainment of a given set of ends.[66] He discusses cost-benefit analysis criticizing the fact that intangibles do not really enter into the analysis, so that the effects measured in money terms are treated as the most important effects. It may be the case that intangible effects outweigh the tangible effects. He also discusses Litchfield's "balance sheet of development," which considers all benefits and costs with respect to all the community goals in one enumeration. However, the costs and benefits enumerated refer to different objectives and are not all relevant for a single objective. Benefits and costs have meaning only in relation to a well-defined objective, and thus the maximization of net benefits in the abstract is meaningless.

Distinguishing between goals (a goal is an end toward which a planned course of action is directed), objectives, and policies, Hill specifies that for his goal-achievement matrix goals should be defined operationally as objectives. Requisites of the system are the necessary conditions (but not the sufficient conditions) such as feasibility, immediacy, and interdependence. Consequences that are positively valued are benefits; those negatively valued are costs. This means that Hill's analysis does not consider opportunity costs of the project with respect to alternative investment. Given the ordering of community goals, and alternative plans to meet these goals, the task is to identify the plan which best satisfies the community's goals. Weighting is introduced to reflect the community's valuation of various objectives and their evaluation of the incidence of benefits and costs. These relative weights might be determined by asking decision makers to weigh objectives, by general referendum, by an interview sample of persons in the affected groups, by identifying the community power structure and asking their views on weighting, by public hearings, or by analyzing the pattern of previous allocation of public investment. Hill feels that uncertainty should also be considered, at least indirectly through use of conservative estimates, requirement of safety margin, and so on. Time preference is reflected in the rate of discount, which represents the opportunity cost of deferred consumption applied to costs and benefits over time reduced to present value.

Hill classifies objectives into those mainly affecting users of transport facilities, those which principally affect the immediate environment of the transportation route, and those mainly affecting the entire urbanized area. For example, the objective "reduction of community disruption" might be defined as the direct effects on communities, adjacent to the proposed transportation improvement resulting from route location. The effect on community disruption could then be measured by (1) displacement of residential, commercial, industrial, and institutional buildings, and (2) boundary effects. Displacement perhaps should be minimized, since financial compensation may be less than the actual money loss. For example, the psychological costs are not compensated for. Displacement is harder on older residents and on groups facing discrimination. A highway might reinforce boundaries between two neighborhoods, or set up barriers in a once-homogeneous community.

Hill says: "By determining how various objectives will be affected by proposed plans, the goals-achievement matrix can determine the extent to which certain specific standards are being met."[67] The key to the method is the weighting of objectives, activities, locations, and groups or sectors in an urban area, leading to a unique conclusion. In this sense the model resembles the scoring model set up by Schimpeler and Grecco. This modification of cost-benefit analysis has advantages since it attempts (but often fails) to measure many of the intangibles. The tool is limited since, like cost-benefit analysis and the balance sheet of development, it cannot determine whether a project should be executed or not. The need for the proposed project is assumed to be given, and the method aids choice between alternative plans. Another major drawback is that interaction and interdependence between objectives is not registered in the analysis, so that the method is more useful for planning in a single sector rather than for multisector planning. The method is very complex, time-consuming, and suffers from a lack of data availability.

A weakness of these models, and of contemporary economic theory, is the assumption that preference scales are stable and determinant. Preferences are formed through an interaction between individuals, households, and the society at large. Fitch says that the substitute for the market process in the social arena is the goal-making process.[68] He believes that setting up goals and gaining popular support for them is the major role of political leadership. The authors discussed have set up theoretical constructs which attempt to account for the intangible social and economic aspects as well as considering the more traditional, easily quantifiable factors. One of the major disadvantages of these methods is that not enough is known about how to determine and rank community goals. Another major consideration should be the type of decision-making body involved. If the intangible effects can eventually be quantified in an acceptable way and if these effects are then found to be of minor importance, planners will be justified in using the simpler models which include only quantifiable factors. However, it may be the case that these intangible factors include the most important effects from the welfare point of view. If so, they certainly should not be neglected.

An attempt at combining cost-benefit analysis with the matrix approach to transportation decision making is offered by Kuhn who views the economy in an accounting framework, where "costs" in one account must be balanced by "gains" in another.[69] After setting up broad goals for public enterprise economics, Kuhn compares a subway with an expressway in a model balance sheet, showing all traceable costs for each project. This method includes intangibles, so that he must assume that the different kinds of values can be reconciled in quantifiable terms. The solution is then found by providing service to the point of intersection of the marginal cost and marginal gain curves (which are derived from the balance sheet). Since project selection occurs over time, interest rates, possible interrelationships of projects, and budget constraints must be considered. Quantifying nonmarket costs and gains is very difficult even when the decision maker is concentrating on future investment rather than on present pricing policies. Kuhn includes all system effects internally in his analysis, which means that the model has a high number of nonmarket values. Market imperfections and the time dimension make cost-benefit or uniform pricing plans operate in ways not predicted by theory.

In replying to Kuhn, Vickrey feels that nonmarket values are insignificant or irrelevant. This would reduce measurement problems.[70] Otherwise, he could not advocate his peak-load pricing scheme. Thompson, however, agrees with Kuhn as to the importance of nonmarket values.[71] Until this aspect of the pricing controversy has been settled, it seems important to continue the attempt to quantify intangibles. Once this measurement has been accomplished, if nonmarket effects are indeed insignificant and the distribution of the burden is not a problem, then planners will be justified in dismissing them. At this point, however, it seems unlikely that intangibles are unimportant.

1.5 Formal Transportation-Interaction Models

As transportation theory and urban economics developed, it became increasingly clear that transportation and community structure act upon each other, rather than independently. Thus, a number of the larger transportation studies began to explore interaction models to explain city structure and to predict growth of urban areas and transportation needs.

In general, there are two views of the role of transportation. Planners often feel that the provision of transportation facilities is a crucial factor determining the form of the urban area. Therefore, all urban problems and values should be considered. Others, such as Meyer, Kain, and Wohl, feel that transportation is a less crucial factor, and they emphasize labor needs and space requirements in using location theory to explain the shape of the city.[72]

Emerging metropolitan policy issues are concerned with the costs, benefits, and incidence of public and/or private investment as affected by public action. Steger and Lakshmanan proposes that the impacts of public investment be studied from regional, locational, and interregional views, according to a "geographic hierarchy."[73] Impact modeling strategy involves trade-offs between: the diversity of impact types to be estimated, the dimensions of impact incidence (areal, sectoral, population class, temporal), the state of the art of analytical techniques and the richness of the information base. Steger says, "No single analytical method, however, is likely to accomplish the forecasting requirements for evaluation of urban/regional public investments."[74] For improved evaluation, he recommends considering spatial allocation as a factor of production in the same way that capital and labor are considered at present.

In planning for public investment, the crucial factors are who builds the systems, who pays for them, and who receives special benefits. In his comments on Steger's paper, Lyle C. Fitch says: "The essence of planning is coping with, and inventing ways of handling, uncertainty."[75] This is particularly true in a large interaction model, where specification may not be satisfactory and where the large amounts of required data introduce inaccuracies. Thus, it may be more useful to choose a range for uncertain variables, and also to choose carefully among the alternatives to analyze, in order to preserve flexibility in the planning process.

Better understanding of the processes relevant to policy decision can be obtained through the use of computerized techniques for projection of the impacts of specific proposals. Specification of development goals is of paramount importance. An operational model can generate estimates of employment, income, or population for any given level of the controllable independent variables (i.e., any given policy proposal). Levin examines regional goals with respect to optimization.[76] Using a linear programming format, he sets up a model to maximize regional output (as one possible goal):

Maximize $$V_p = \sum_{i=1}^{n} p_i X_i$$

Subject to $$X_i = \sum_{j=1}^{n} a_{ij} X_j + a_{io} X_o \quad (i=1, \ldots, n)$$

$$X_i \geqslant 0 \ (i=1, \ldots, n)$$

and $$X_o \leqslant K$$

where

V_p = net output or value added in production in the region

X_i = output of the ith industry

p_i = value added per unit of output of the ith industry

a_{ij} = amount of the ith industry's output needed for the production of one unit of output of industry j

X_o = total labor input, and

K = available labor supply

Objections may be raised to this method, since linear production functions may not be realistic and coefficients may not be stable. Maximization of employment opportunities is important, but planners should also consider the level of earned income per gainfully employed individual, wages, stability and growth of income and employment over time, the ratio of wages to commodity prices, and the distribution of income.

Coordination with larger units is essential, since optimization at the national level may be inconsistent with regional optimization. This may occur because:

1. With multidimensional goals, regional preference patterns may be such that there will not be a consistent ordering of these goals at the national level
2. Even if maximization output were the only goal, each region's share in the relocation costs needed to achieve it may not equal their share in the resulting gains
3. External economies or diseconomies extending across regional boundaries have not been taken into account

Improved regional planning policies depend on:

1. The relationship between population mobility and economic efficiency
2. Comparison of economic well-being for different individuals
3. Priority of national over regional goals
4. The extent to which an individual's well-being depends on the welfare of his community, independent of his own situation

In setting up a model for multiple planning goals, not all the factors potentially relevant to economic and social welfare can be taken into account because of the large number of such factors.

The Penn-Jersey Transportation Study was designed to assist planners in promoting a desired pattern of regional development as well as to provide for transportation needs.[77] On the other hand, RAND initially viewed the transportation system as a response to growth in an area, rather than as a causative factor of growth.[78] At the time when interaction models were first set up (early 1960s) there was disagreement about the spatial organization of the urban community, with confusion over what it "should" be like versus what it "will" be like. Early studies were valuable primarily because of their estimation of the scale and directional distribution of future trips.

In an attempt to better understand the relationships between transportation and the spatial organization of economic activities, RAND worked with three basic land occupancy sectors: industrial, commercial, and residential.[79] In estimating the value of land surfaces, RAND recognized constraints such as the racial problem, and developed an econometric model, based on relationships estimated from cross-sectional data. The model is useful for structural estimation and diagnosis, but it is not a forecasting model. "The model is intended as a vehicle for evaluating the general underlying technological and economic changes occurring in today's urban areas."[80] Kain feels that the importance of secondary effects has generally been overestimated, but that any partial analysis should postulate the nature and extent of these effects.[81] The major variable for urban transportation planning in the work done by RAND is the journey-to-work trip by urban residents.

Kain, concerned with the locational choice of a household, hypothesized that households substitute journey-to-work expenditures for site expenditures, with the substitution dependent largely on the household preference for low-density as opposed to high-density residential services.[82] He made five key assumptions: (1) transport costs rise with increasing distance from the work place; (2) in the market for residential space, the price per amount which a household must pay for a given quality decreases as the distance from the job increases; (3) there is a fixed workplace; (4) households maximize utility; and (5) residential space is not an inferior good. Testing his model with data from the Detroit Area Traffic Study (DATS), Kain found that the commuting pattern was largely from residences in the outer rings to work places in the inner rings. The average journey-to-work decreases with the workplace's distance from the central business district and the proportion of a ring's workers living nearby rose as the workplace was farther from the central business district. Also, workers in higher income occupations and working in inner rings tended to make longer journey-to-work and reside in outer rings. The smallest and largest families make the shortest work trips (perhaps due to lower space preference of small families and a per capita income constraint for large families). Longer trips are generally made by those living in one family units, and women usually work nearer home than men do.

Edin extends Kain's model, saying that the journey-to-work is important in determining residential location, since work-trip expenditure is a large part of the household budget and time is important to the commuter.[83] The commuter is also assumed to consider location rents and the mode of transport in the trade-off between housing expenditures and transportation expenditures. Edin assumes that the family follows a specific behavior sequence: (1) the residential density desired by the household is determined, (2) a decision is made about whether or not to purchase a car, (3) a decision is made about whether to drive to work or to use transit, (4) then the location of the residence is chosen (determining the length of the journey-to-work).

Using data for Skokie, the empirical results were not completely satisfactory due to colinearity among the exogenous variables. However, the model did explain a major share of the variation in behavior based on quantifiable economic and social variables, indicating that "sentiment" is probably not a major factor in residential location decisions.

In this same vein, Moses set up a model to show what wage would have to be offered at any given location in order to attract labor from other places in the urban area.[84] He assumed that the area starts from a utility-maximizing equilibrium, where residents have already located according to the trade-off between residential location and transport costs. Then a firm wants to attract more workers, and does this by offering higher wages, inducing workers to travel farther. However, his model does not consider the actual distribution of population, and it treats the wage in the city core as given rather than as determined.

Using Pittsburgh data, Lowry set up a computer model of the spatial organization of human activities within an urban area.[85] His model is designed to be used for: evaluating the impacts of public decisions on the urban form, and predicting the changes in urban form over time as a result of changes in the key variables (i.e., the pattern of basic employment, the efficiency of the transportation system, and population growth). The location model is a set of simultaneous equations whose solution gives an equilibrium for land use patterns and the distribution of employment and population, which is a rough simulation of the actual behavior of firms and households. In the iterative solution, Lowry finds that the impact of local changes tend to be localized. The system suffers from the fact that even if it does have a set of solution values, they may not be unique.

Lowry sets up a model for the market of urban land, which he then uses to classify some of the major urban development models.[86] Since willing buyers and sellers operate in the land market, site transactions will determine the spatial organization of activities. Sites are held by owners, and potential buyers are referred to as establishments. At the start of each transaction period, each establishment appraises the merits of every site, deciding what price it would pay for each (so that it would be indifferent among locations). Figure 1-5 shows these demand prices as a matrix. If the matrices are published, site owners can scan their column to identify the tenant willing to pay the highest price, so that competition among potential occupants determines the market price. This view is oversimplified, since the actual market operates by bid and counter-bid, and there is no "transaction period." However, demand price in general may be represented by

$$p^{hi} = f(X_1^h, X_2^h, \ldots ; Y_1^i, Y_2^i, \ldots ; Z^{hi}, h=1, \ldots n)$$

Establishments	1	2	3	4	5	6	7	8
1	3	7	16	10	10	15	20	5
2	9	4	9	12	2	8	13	7
3	5	8	15	6	3	7	11	19
4	7	9	9	17	8	14	19	16
5	2	6	16	10	15	17	18	5
6	8	7	11	16	14	18	3	9
7	6	8	17	13	9	3	17	14
8	3	10	14	7	5	15	2	10

Where the number equals the demand price for the site,
(hatched) indicates the initial location
(box) indicates the terminal location.

Figure 1-5. The Market for Urban Land. Source: Ira S. Lowry, "Seven Models of Urban Development: A Structural Comparison," in *Urban Development Models,* Special Report 97 (Washington: Highway Research Board, 1968), p. 124.

where

h is a particular establishment,

i is a particular site

X_1^h represents the establishment characteristics

Y_1^i represents the site

$Z^{hi}, h\text{-}1, \ldots, n$ represents the site location with respect to the location of other establishments.[87]

The relevant X's, Y's, and Z's are difficult to identify and measure. Moving costs tend to give stability to the model so that few establishments move in any one period.

Since site characteristics may be modified, there may be an investment function by which owners evaluate site improvements:

$$E_j^i = g(C_j^i, P_j)$$

where

i = a specific site,

j = a specific bundle of site characteristics,

E_j^i = expected gain from converting site i to condition j,

C_j^i = expected cost of conversion,

P_j = current market price of sites in condition j.[88]

This introduction of uncertainty has given the model a dynamic form. For empirical purposes, it is useful to group sites into "districts" and establishments into categories (such as household, industry and commerce, government) to give a reduced matrix. By comparing initial and terminal land use distributions, land use succession is shown in the columns, and migration patterns are shown in the rows of the matrix.

Lowry describes the Chicago Area Transportation Study (CATS) land use model as a column model. It has an inventory of land uses projected for each district separately, based on the district's initial inventory, its zoning map, and its location. The information does not include comparisons of districts with respect to their desirability as a location for establishments belonging to a given activity group. The model includes an account of the land use in each district, but does not include origins of new tenants or destination of those who leave. One is not able to say what structure of demand prices is consistent with a solution of the CATS model, or anything about the evaluation and investment factors which motivate establishments.

The University of North Carolina (UNC) Model is used to predict the incidence of conversion of rural or vacant land to residential use as population grows. The model assigns an "attractiveness score" to vacant land "cells," and the probability of conversion of a cell is assumed to be proportional to the attractiveness score. Behavior of land developers is not specifically reflected in the UNC model. Land use succession is determined by demand rather than by entrepreneurial decision. The model does not include establishment characteristics (the *X*'s), only site and accessibility characteristics (the *Y*'s and *Z*'s). As with the CATS model, nothing can be said about the origin or destination of movers.

The EMPIRIC location model was designed by Traffic Research Corporation for the Boston Regional Planning Project. It reallocates population and employment among the region's subdivisions as regional totals change over time, and as locational changes occur in the quality of public services and transportation networks. Using simultaneous equations, the model deals with two classes of population (blue collar and white collar) and three classes of employment (wholesale and retail, manufacturing, and all other). The emphasis is on the change-in-share of each district, with little application of land use accounting. This is a row model, with only net changes showing up. There does not seem to be a market interpretation of this allocation of an activity among districts.

The POLIMETRIC migration model (also set up by the Traffic Research Corporation) is a set of simultaneous nonlinear differential equations, one for each activity in each district. The dependent variable is the rate of change of the level or volume of the activity. Redistribution of activities occurs only through interdistrict migration. Land use accounting is suppressed (a row model). Interdistrict movements stem from comparisons of the merits of alternative locations for each activity. The model has no market interpretation, since it merely shifts activities from low desirability districts to above average desirability districts.

Lowry considers the Pittsburgh Model (his *Model of Metropolis*) to be a hybrid since it includes a stronger system of land use accounts than either the EMPIRIC or the POLIMETRIC model. Three classes of employment and one of population are allocated along tracts, subject to a maximum density constraint. Tracts are scored by their accessibility. The model suppresses land use succession and internal migration. Residential density increases with increased accessibility of a tract to places of employment, and decreases with increasing amount of space available.

The Penn-Jersey model is based on market demand for residential land. The landowner function is limited to choosing among potential tenants, and all entrepreneurial behavior is suppressed. The solution is an equilibrium allocation of households to residential sites such that the "rent-paying ability" of the region's population is maximized. The data requirements for the Penn-Jersey model are far greater than those for the CATS, UNC, EMPIRIC, POLIMETRIC, or Pittsburgh models. Limits of computer storage capacity can become a problem with this approach. The Penn-Jersey model is not designed to study patterns of population movements or land use succession.

A.D. Little, Inc. developed the San Francisco model that emphasizes market supply. Given biennial forecasts of construction and demolition, changes in the physical condition of structures, rent levels, and occupancy rates, the model is designed to analyze the impact of various public programs on the housing stock and its use. The model uses a detailed breakdown of dwelling units and residential population in assigning the population to available housing. The household selects the dwelling (desiring to pay minimum rent) rather than the owner selecting the tenant. Thus, the model does not necessarily lead to a market-clearing solution. If the market does not clear, this shows the need for changes in the housing stock, either through building or through demolition. The solution of the model is then the housing stock inventory and pattern of rents after the changes take place. The San Francisco model is limited since it does not consider accessibility, but it does have rigorous land use accounts.

The major problem with this type of analysis is in making a theoretical model usable with the limited data available. Lowry concludes that the CATS model is acceptable if the planner only desires land use forecasting at a level of detail adequate for transportation planning. However, the CATS model does not aid in understanding the spatial organization of the city, nor in evaluating alternative land uses that might be achieved through public policy. However, although not a forecasting tool, the San Francisco model is helpful for analyzing public policies under varying assumptions. In general, better development of plan evaluation criteria seems to be essential.

Bruck et al., have been concerned with impact modeling and problems of developing measures to evaluate indirect consequences of changes in the transportation network.[89] They attempt to estimate the impact of the transportation system design on projected levels of employment, population, and land use. They include a consistency check with alternative regional spatial designs and regional development objectives.

For impact studies, the outputs of the interregional-interindustry, input-output model are used as inputs in the Intraregional Allocation Model. Evaluation of alternatives then requires identification and measurement of the effect variables and creation of a set of "evaluation accounts" for each alternative investigated. Evaluation is achieved through a strategy of choosing preferred solutions from possible alternatives and converting cost-benefit data about preferred solutions into program budgeting information. The sequential process is shown in figure 1-6.

The Bruck, et al., study is an attempt to use computers for aid in evaluation of decision making. It is oriented toward economic changes and does not consider social effects.

MacRae, dissatisfied with the limitations of cost-benefit analysis, developed

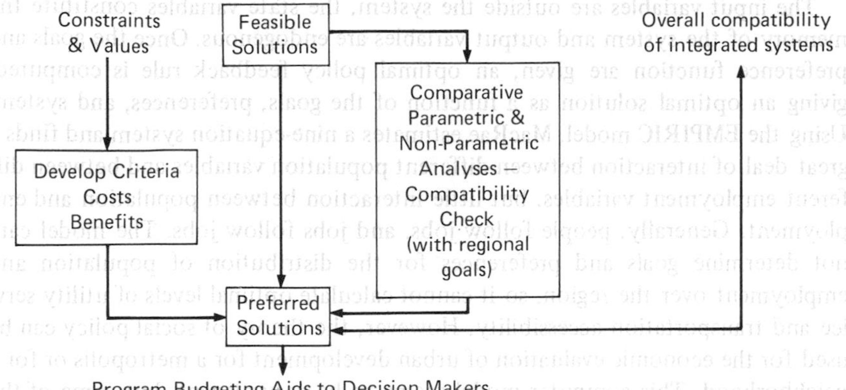

Figure 1-6. Evaluation of Alternative Transportation Proposals. Source: H.W. Bruck, et al., "Evaluation of Alternative Transportation Proposals: The Northeast Corridor," *Journal of the American Institute of Planners* 32 (November 1966); 329.

an alternative which he calls a theory of social policy.[90] He attempts to set up a "computable model of a stochastic dynamic system" which assumes the parameters are known. In the model, current endogenous variables (output variables) are related linearly to policy, exogenous, and random (input) variables and to lagged endogenous variables. MacRae assumes the social goals are given (for the endogenous and policy variables) and that there is a quadratic social preference function on the deviation between goals and actual levels of variables. Therefore, decision makers want to choose policy variables in order to minimize the value of the social preference function (i.e., minimize the difference between goals and performance). MacRae applies his social policy theory using the EMPIRIC location model, which shows the impact of a transportation change on a metropolitan area. It describes the changes in the share of regional population and employment for subregions (traffic zones) as linear functions of past shares of the subregion and of measures of land developability, utilities service, and the accessibility of the subregion. His system is shown in figure 1-7.

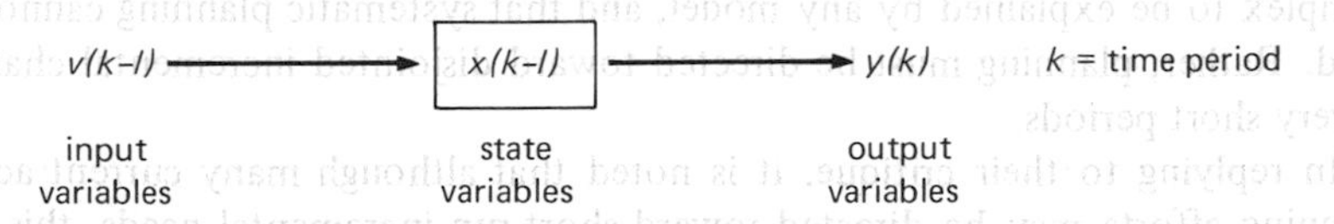

Figure 1-7. System for the Theory of Social Policy. Source: Duncan MacRae, "Economic Evaluation of Urban Development," p. 3.

The input variables are outside the system, the state variables constitute the memory of the system and output variables are endogenous. Once the goals and preference function are given, an optimal policy feedback rule is computed, giving an optimal solution as a function of the goals, preferences, and system. Using the EMPIRIC model, MacRae estimates a nine-equation system and finds a great deal of interaction between different population variables and between different employment variables, but little interaction between population and employment. Generally, people follow jobs, and jobs follow jobs. The model cannot determine goals and preferences for the distribution of population and employment over the region, so it cannot calculate optimal levels of utility service and transportation accessibility. However, the theory of social policy can be used for the economic evaluation of urban development for a metropolis or for a neighborhood. This computer model seems to be very elegant, but some of the data needs are difficult to meet, and the goals must be very carefully specified in order for the results to be meaningful.

Quandt and Baumal attack the competing alternatives problem from a different viewpoint.[91] They define an abstract transport mode by a set of characteristics including speed, fare, frequency of service, and comfort. Then the decision to travel by mode k is a function of the absolute levels of characteristics of the best modes on each criterion, and the performance level of mode k on each criterion relative to the best mode.

Gronau and Alcaly argue that this model fails to provide a better explanation for the demand to travel by mode k than does a simple gravity model of total traffic volume.[92] Transport decisions depend on the mode to use for each trip and on the number of times to travel to each destination. If the total cost, P, is assumed to equal the fare, C, plus the opportunity cost of time mH (where m is the value the consumer places on his time, H is the time elapsed) then $P = C + mH$. This is hard to generalize for more than two modes. An aggregate demand function for travel by mode k, (T_k), might look like:

$$T_k = f(C_B, H_B, C_k, H_k, C_k/C_B, H_k/H_B)$$

where subscript B is for the "best" mode, subscript k is for mode k. However, this ignores m, and addition of new modes may alter the demand structure.

According to the Lindbloom-Braybrooke thesis, much of the preceding analysis on formal interaction models is useless.[93] They argue that the world is too complex to be explained by any model, and that systematic planning cannot be used. Rather, planning must be directed toward disjointed incremental changes in very short periods.

In replying to their critique, it is noted that although many current actual planning efforts may be directed toward short-run incremental needs, this certainly is not always the case. The world has many changes that are much more than marginal, and there may be an overriding system which facilitates their description. Thus, it does seem to be useful to develop interaction models in an attempt to explain and predict a wider range of variables which do not operate independently. In many respects, models that contain transportation sensitive elements hold out the greatest hope for transport planning and project evaluation.

1.6 Systems Analysis

The final approach to project evaluation concerns the total systems engineering attack on planning. Simply stated, systems analysis takes the broadest possible view toward transport evaluation. It is aimed at comparing heterogeneous alternatives in such a way that the decision maker is presented with as much useful information as possible. The results of a systems evaluation are then to be applied by the planner in a supporting manner, since in most cases the actual decision must be made by him utilizing his past knowledge and experience. The systems approach is forthright in its admission that the final decision must rest on elements of subjective human evaluation.

The approach was initially developed by the Department of Defense in the period after World War II. It was instigated due to the growing complexity of the weapons that required careful evaluation in terms of their comparative value. The crucial difficulty with an item by item approach to evaluation was that although the individual parts were capable of passing a performance test, when the entire weapons system was assembled, it often failed to function to desired specifications. To combat this difficulty it was decided that what was needed was an evaluation of the weapons system as a whole rather than the individual components.[94]

The similarity of this problem of evaluating weapons systems to that of transport network planning is evident. Any transport network can be broken down into individual parts and analyzed as such, or it can be treated as a whole so that its overall effectiveness can be seen. If the network itself is approved as a unit in terms of both the relative demand for it and its effectiveness, then it can be assumed that the individual parts are justified as a necessary component of the whole. As a result of this point of view, the highway decision maker would evaluate a plan on the basis of the overall flow of traffic and community impact, using as broad a definition of the area as possible. In this way the emphasis rests on the interrelationship of the parts and their ability to perform the desired tasks. Design and planning must then consider the role to be played by the parts, in terms of both costs and benefits, with respect to the effectiveness of the entire system.

In order to further clarify the relationship between the systems approach and transport planning, it is useful to elaborate on the key technical terms and relate them directly to transportation.[95] First, a *system* is an integrated whole which functions as a single unit to perform a specified set of "activities" so that some recognized "goal" can be achieved. In highway planning this "system" could include anything as large as a nationwide network (such as the Interstate System) or as small as a state or locally financed road program. The set of "activities" performed by the system might involve facilitating the flow of commuter or goods traffic. Finally, the "goals" of the system can be the stimulation of area growth, the development of interregional communication or an increase in the level of national security. It should be noted at this point that these three aspects of the evaluation plan, the system, the activities, and the goals must be defined explicitly for each problem. This is necessary so that the individual objectives and standards of overall effectiveness can be developed. The objectives become the physical translation of the generalized goals while the criteria are the means by which the extent of objective achievement is measured.[96]

Secondly, a *part* is a subcomponent of the whole that employs scarce resources in the performance of some given function. In highway planning this would correspond to a particular road or road segment designed to move people and goods from one terminal to another. The resources employed in this operation would include the fixed materials of construction and those variable inputs required of the road users.

Thirdly, it should be noted that the system under consideration is itself only part of a larger *supersystem* with which it interacts. Such a supersystem involves the *physical* and *social* environment that surrounds any transport network. To disregard the effect of a highway network on these other systems and the repercussions that they in turn force onto the highway can lead to serious planning errors. For instance, as the science of man and his environment progresses, the ecological effects of transport planning become more apparent. The impacts of pollutants on the environment, the use of open space and conservation land for construction purposes, and the introduction of new forms of technical design have all brought about noticeable changes in the natural world in which the highway is located. In turn, this natural world is capable of reacting, through long-run changes, on the highway system itself by causing either an alteration in system goals or functions.

With respect to the quality of the social atmosphere that surrounds the highway, this too can be altered by the system, especially in densely populated urban areas. In these regions, the construction of major arterial facilities can create neighborhood instability, personal disruption, and a general decline in urban living conditions. Depending upon their magnitude, these social impacts can in turn affect the political atmosphere which itself is capable of generating corrective forces aimed at the planning of the highway system. Economic impacts also enter into the interaction scheme as well. By generating unanticipated industrial development, the highway network may be taxed beyond its planned capacity. On the other hand, if it by-passes or somehow destroys a key source of area industrial strength, then this may lead to underutilization of the facility. Obviously, these possible supersystem effects must be considered in the decision-making process. Only by taking the broad systems approach can this task be adequately accomplished.

In implementing the systems analysis technique, there are three phases of operation which must be undertaken.[97] The first of these has already been discussed in some detail and requires the development of a set of goals and alternative systems which are designed to achieve these goals. The purpose of the analysis is, then, to determine which system is most *effective* relative to its *cost* in achieving the goals within given constraints. Hence, the second or *cost-effectiveness* phase of the analysis is born. Before passing on to the discussion of this second phase, it is important to emphasize two essential points in phase one. First, in transport planning the goals should not be solely transport oriented, but rather they should reflect the broader community point of view. This is due to the previously noted interaction of the transport system with the surrounding supersystem. The second point is that all modes of transport should be investigated in order to view the entire spectrum of alternatives which are capable of achieving the desired goals. This might even include an analysis of technical forms, such as those which substitute communications for transportation.[98]

The second or cost-effectiveness phase involves both the development of measurement criteria by which to judge the performance of each alternative and the establishment of the fixed boundary of system cost or effectiveness. In developing measurement criteria, several aspects of transport system evaluation are susceptible to direct quantification. These would include driving time, ton-miles, passenger-miles, fuel expenditures, and the like. Others such as safety, social stability, aesthetics, and attitudes towards the governing body elude easy measurement. However, given the particular situation, all or some of these might be accepted as valid criteria by which to judge the effectiveness of a system. What is needed here is a set of criteria based on the goals or objectives that are deemed relevant for the specific problem and which can be used to measure and test the systems performance characteristics.

The establishment of cost or efficiency boundaries highlights one of the central problems which must be faced in practical resource allocation, namely that in plan evaluation there is a choice to be made between "economy" and "efficiency." In the final result, the one should be the mirror image of the other. However, in planning it is necessary to hold one aspect fixed and use the other as the target. In particular, it must be decided which of the two following approaches is to be followed:

1. Enforce a strong budget constraint which limits the amount of money that can be spent on developing the system. With this constraint, the objective becomes choosing the system that comes closest to maximizing the achievement of the goals.
2. Enforce a fixed level of system effectiveness which specifies how close the

system must come to satisfying the goals. With this constraint the objective is to choose the system that minimizes the budgetary outlay.

In the area of highway planning the second approach is usually adopted, since in many cases the facility is constructed in response to an existing or projected level of consumer demand. Consequently, the budgetary allocation becomes the variable factor used to achieve the desired effectiveness. This approach, however, should not be considered as the single best choice of the two in all situations. Rather, its use must be tempered with rational judgment. This is due to the limitations that can arise in cases where the costs are not obvious or are not easily quantified.

The third and final phase of the systems approach is that of determining and comparing the relative capabilities and effects of the alternatives. This requires an actual projection and presentation of the performance of each alternative. The usual (but not mandatory) procedure followed in this stage is to develop a formal model (mathematical or otherwise) which describes the operation of the system. Such a model can relate the dependent variable of effectiveness or cost to the independent performance characteristics that vary between systems. Hopefully, given the constraint that has been adopted, a general level of technical similarity between systems and a relatively small number of "unimportant" nonquantifiables, the results of the model should indicate the preferred system. However, if the impacts, costs, and technology of the alternatives are widely divergent, then the stability and value of the model come into question. As a result, a substitute evaluation approach must be found.

In many cases a matrix presentation is adopted that is similar to the ones already outlined (see section 1.4). The usual approach is to include all relevant items whether quantifiable or not. The alternatives are then compared on the basis of how well they fulfill the criteria in general and the most important ones in particular. When the impact of the different alternatives falls on various criteria which are not affected in the same way by each system, the choice between programs must be made by the decision maker. Given his background and experience, he should be able to apply a subjective sorting technique to the factual material and select the desired system. The benefit of this approach is twofold: First, it indicates explicitly that the final choice involved subjective human evaluation which took into account the nonquantified impacts that are usually omitted from other approaches. Second, it clearly outlines both the problem at hand and the objective data employed to guide the final decision. In providing this information, others not as closely connected with the topic can review the reasoning that brought about the decision.

The final aspect of this third phase is the application of sensitivity analysis to some of the data inputs that are capable of variation under different surrounding conditions. To account for this element of uncertainty it is often wise to provide an alternative set of reasonable data or assumptions which are also subject to cost-effectiveness evaluation. The results of this final step will show the extent to which the decision is subject to change if crucial elements are allowed to alter. If serious changes in system choice are detected, then a more detailed study of the alternatives may be warranted. At least, those in charge will have a complete picture of the possible consequences that the assumptions can have on the system selected.[99]

In applying the technique of systems analysis to the area of highway planning, it is necessary to describe some of the more important limitations that can be encountered.[100] The first of these involves the human subjective element. The final decision is itself only as good as the experience and judgment that went into making it. Great skill and professional ability are required in the cost-effectiveness evaluation and in the ultimate selection. This necessitates a dual capability in both the general *technique* of the systems approach and in the specific *area* of application. A weakness in either field reduces the probability that the correct choice will be made, because such a weakness can lead to the temptation to adopt paths of logic that are deceptively false. For instance, the use of functional relationships which are assumed to be held constant between systems is one source of difficulty. Another logical difficulty involves the improper use of probabilities of achievement or the introduction of the risk of system failure. In transport planning this might involve the implicit assumptions of concurrent or future area development projects which if they occur will justify the use of a particular system. However, what must be introduced is some measure of the uncertainty attached to the outside effects and the element of risk that this adds to the selection process. The decision maker must have the ability to evaluate the reasonableness of these risk factors and to view their potential effects.

A second source of difficulty involves the process by which the criteria used to evaluate each system are chosen. Here, the directions of error are several. First, there is the tendency to select those criteria which possess a recognizably measurable dimension and to skip over those which are hard to quantify. As we have seen in other parts of this survey (see sections 1.1 and 1.4), this difficulty is found in other evaluative frameworks. To correct for this, it should be the purpose or goals of the system not the extent of measurement that dictates the criteria.

Second, there is the tendency to select too many criteria. This generally results in confusing the view of the relative effectiveness of each system. If as the number of criteria expand, the proficiency of previously ineffective systems group as they excel in the new dimensions, then it becomes necessary to specify which criteria are most important. As a result, the final decision requires the elimination of criteria which themselves are deemed relatively unimportant. Obviously, the complications introduced by this multiplicity of criteria can be extensive, especially if there is more than one evaluator.

The final source of difficulty with criteria is introduced if, instead of an overabundance of criteria, there are too few. The selection of one or two ways by which a system can be evaluated can give a disastrously false impression of its potential impact. An example of the application of this problem to the area of highway planning arises where the planners focus on only one aspect of a networks performance. Since one obvious purpose is to facilitate the flow of traffic, alternative systems can be erroneously evaluated solely on the basis of their ability to minimize the dollar costs associated with that flow. In limiting the number of criteria in this way, the external impact of the facility on the community is overlooked especially since such impacts are hard to put in dollar terms. It is highly unlikely that such a one-dimensional test of effectiveness will be sensitive enough to allow selection of the best system.

The final limitation of the systems approach as it is applied to highway planning concerns the extent to which the evaluation of the entire system relates to the route selection of individual parts. The selection of an overall network plan, as noted previously, justifies the construction of individual segments. However, the exact location of the road part has not been determined, and by varying the site there can be significant fluctuations in local and regional effects. A repetition of the analysis may have to be undertaken on the community level to determine the specific ground route. Here again the question of who should set the goals and how they are to be chosen must be decided first. This then raises the possibility that a conflict between local and area-wide goals can arise. Furthermore, the cost in both time and money may restrict the scope of the local study, thereby limiting its value. Finally, the technical competence of local decision makers may also reduce the reliability of the results.

In concluding the review of systems analysis, it seems quite clear that the main strength of the approach lies in its ability to evaluate broad questions of resource allocation relative to specific goals. In doing this, it is able to test the implications of these goals against widely diverse solutions.[101] The results of such tests become the background upon which the final decision is made. Its principal limitation falls in the area of individual project evaluation where the micro-oriented nature of the decisions which must be made severely tries the general nature of the approach. Given that the task of an evaluation may be to judge the relative worth of alternative locations for a specific part of the project, it is difficult to do this within the systems framework. What is still required is a systematic procedure for evaluating the costs and benefits of the individual parts once the overall plan has been approved.

2 Background for the Model

2.1 Introduction

Given the literature survey as a starting point, chapter 2 contains a description of the generalized characteristics that must be a part of any transportation investment model. It is intended that these characteristics define certain qualities of the model and establish its operational boundaries which will reveal both the strengths and weaknesses for the investment decision-making process.

In developing the model two guidelines were followed. First, as noted earlier, it is designed to both identify and quantify the social, economic, and environmental factors that are relevant to a given corridor location decision. In addition, it must be possible to link this information to the engineering data in order to select the optimum alternative according to the appropriate criterion. This practical aspect lies at the core of the analysis. The model must maintain a strong contact with reality which will allow for its direct and repeated application by properly trained personnel.

A corollary to this consideration is that the model should also be efficient in the economic sense with respect to the internal use of resources. The time and money spent upon the implementation in specific uses should permit the achievement of the fixed objective at a minimum cost to the user. Any tests that measure the degree to which this efficiency condition has been met must focus upon a comparison of the incremental costs and gains associated with the application of the various analytical approaches that claim to achieve similar results.

The second guideline was that the model should have anticipatory capabilities. The benefits of this requirement are evident when the nature of the planning process itself is considered. After a highway is built, as an illustration, its impact on an area can be most easily assessed. For example, a "before and after" study gauges the extent of social and economic change within a community. Residents are vocal in their description of the major realized alterations that are of prime concern to them. Also, after construction, the value of the facility to nonusers as well as users can be determined.[1] In some instances, these after-the-fact impact reviews indicate the existence of severe problems or unfavorable effects that were not considered before the construction took place. These errors of omission are not usually brought about by careless planning. Rather, they appear as a result of the lack of current information regarding either the nature of the region or the preconstruction aspects of the community that are considered to be most important. Such effects as residential fragmentation or the disruption of previously favorable environmental conditions are key changes that appear only after the erection of the facility. What is needed is a means by which these effects can be anticipated during the planning phase of highway location.

To meet this requirement, the potential for and relative importance of each impact must be brought to the attention of the transport planner. If an investment model achieves this end, then the energies of the planner can be oriented toward the most crucial effects. Obviously, cost limitations will not permit the same in depth evaluation of all the possible effects of a highway. However, if the range of an investigation can be narrowed, for example by having the model establish a hierarchy of impacts, this will allow a focusing of the review activity. In addition to indicating those impacts which are considered to be vital if they do occur, this evaluative technique would provide the planner with information regarding the subjective attitudes of residents with respect to the potential location of a highway corridor. In this way, the model performs an anticipatory or predictive function.

In the following sections characteristics considered to be relevant to the model are presented. These discussions include not only a review of a particular attribute, but some also contain an analysis of the alternative ways by which it can be accounted for. In general, these reviews are a mixture of descriptive statements of fact, perspectives that are being utilized, and choices of methodology which must be considered both in the construction of the model and in the planning process itself.

2.2 Generalized Attributes of a Transportation Investment Model

By defining the points discussed in the following pages as attributes, the intention is to provide an indication of the character of the model. Typically in the physical world, an attribute describes either a tangible aspect or a character trait that helps to distinguish the nature of the item under review. For example, a person is tall or short, fat or slim, with red hair and green eyes or some such combination. In addition, he is loyal and trustworthy, or shifty and dishonest. Upon completion of the list, an onlooker has a mental picture of what the man is like. Any unanswered questions with respect to the "soul" of the man would, however, have to wait a personal review of the subject in action. The same can be said regarding the attributes of the model. They assist in the recognition and interpretation of the model, but any formal understanding of its internal operation requires a personal encounter.

The model itself is aimed at the evaluation of a transport system. However, the Systems Analysis discussion in section 1.6 demonstrated that a transport system does not exist or function apart from the much larger world that surrounds it. This concept of an integration of apparently independent systems or supersystem necessitates that the interaction between the social, physical, and transport networks be considered by an evaluative process. Too often the development of a transportation system is assigned a passive role by assuming that it is being designed solely in response to already existing or projected user requirements. No recognition is taken of the fact that as a sizable fixed investment in social overhead capital, the network is capable of reflecting back upon and altering the environment in which it exists. Indeed this interaction is likely. For example, the physical or natural environment is affected by the ecological impact of a structure.[a] This can occur by the filling of marsh lands or the intrusion of the facility into land that was previously used for wildlife preserves or other conservation purposes. Obviously, such effects, if severe enough, can also cause a reaction in the transport network through changes in both design and location.

The same is true of the social or human environment. Although the network may be originally designed to facilitate the flow of men and goods, it should not be overlooked that it can also affect the extent and form of social relationships between people. The number and direction of personal or business contacts and the quality of the social services provided to the community are both capable of being altered by highways. Such alterations, plus the hardship caused by physical dislocation can create tension within the community.

Obviously this active role of a highway should be considered in the planning and design phases. In focusing on this interaction however, the system effects need not be limited to those active impacts which are harmful to the community. Rather, there can be significant external effects which are beneficial to the growth and development of areas along the right of way.[b] For example, the physical environment may be improved if the construction of a limited access highway draws existing through-traffic off crowded urban streets and thus reduces the level of congestion and pollution in the city. Moreover, the concentration of traffic flows along a new and wider route can facilitate a more advanced attack on noise or exhaust fumes. The building of a new road may also allow the development of scenic or recreational land which was previously inaccessible. This extension of leisure time opportunities is of significant benefit to area residents. To ignore these active effects in the location or design of a network does a disservice to those who are being asked to bear part or all of the dollar burden of highway cost.

In addition to the need for the recognition of system interaction, any attempt at evaluating the potential effects of highway construction must utilize the existing body of knowledge as a starting point. The model should have a distinct lineage, with origins and characteristics that can be traced back to its predecessors. Two broad sources were utilized in developing the state of the arts in the

[a]A timely controversy demonstrating this point surrounds the construction of the cross Florida barge canal.

[b]These are apart from the transferred user effects such as increased site rents due to better mobility or increased access to surrounding locations.

field. The first was the extensive survey of the literature undertaken in sections 1.1-1.6. This allowed for a comparison of current techniques and focused upon the areas requiring improvement. Two of the more obvious areas were adopted as the dual goals of this work. The second information source consisted of personal interviews and consultations with representatives of diverse federal, state, and local planning agencies. In each of these discussions, those being interviewed expressed a sincere interest in any approach that was capable of upgrading the quality of the information in this area. It was recognized by all that the consideration of the external effects of highway use is an essential part of any rational investment decision.

On the federal level, representatives of the Bureau of Public Roads were questioned concerning their interpretation of the social, economic, and environmental section of the Federal Highway Act of 1968.[c] This passage dealt with the consideration of external effects in the planning and evaluation of federally aided projects. In general, their attitude was that PPM 20-8, which detailed the bureau's policy on the 1968 act, was meant only as a general guideline to state authorities in the community impact area.[2] The list of twenty-three items was included as a starting point for state thinking, with the organization, expansion, and operational definition of the terms being left for regional interpretation. Furthermore, the way in which these items were to be considered was not specified.

State officials, while interested in the general aspects of the federal directive, were more concerned with applying these guidelines in actual situations. There were two areas in which they felt practical assistance was required. The first was that of providing a framework whereby the effect of a highway on the community could be viewed. This would involve the identification of the separate effects and a means of considering their distribution throughout the area. The second was the evaluation of the community impacts of alternative routes and designs that would allow for similar traffic flow objectives. To achieve this end, some means of making relative comparisons of the impacts was needed so that trade-off ratios might be utilized.

Finally, on the local level, individual representatives of the established regional planning agencies were interviewed. Since their own efforts had already resulted in considerable progress in the area of regional development, they expressed an interest in further coordinating the efforts of their research with that of federal and state departments of transportation. The unanimous feeling offered by these professional planners whether interviewed in groups or singly was that existing techniques of evaluation placed considerable weight upon the needs of the state while requiring local residents to bear the burden of adjustment caused by highway construction. Too often the trade-offs of user benefits and nonuser costs were ignored in favor of developing interconnected road networks that fulfilled the projected desire lines of travel for hypothetical commuters.

[c]Federal Aid Highway Act 23 U.S.C., Sec. 128(a)(b).

Finally, the unquestioned acceptance of the assumption that highways alone could or should fulfill this anticipated demand further aggravated the regional planners.

The importance of the above sources of information cannot be overemphasized. On the one hand, the literature provides a review of the theoretical structures available in the area plus an indication of the applied work undertaken by others and aimed at solving the problem. On the other hand, in discussing the matter with responsible officials working in the area, an indication was developed of the strengths and weaknesses of the different avenues of attack. Moreover, the stress they placed upon the need for coordination and the integration of available information among governmental levels provided additional support for the development of a broadly-based evaluative model.

2.3 Goals vs. Impacts As a Basis for Evaluation

The first attribute that requires individual attention involves the choice that must be made between a goal or impact orientation for the model. Every model is faced with the problem of abstracting the relevant information from the complex world of reality. The frame of reference or criterion utilized to facilitate this selection process has dominate control over the extent and type of information that the model receives. Consequently, the criterion should be unbiased (which may very well be impossible), or if it is biased, the nature of the distortion should be noted. In addition, if possible the bias should be pointed in a direction that will allow planners to arrange information in its most usable form.

There are two general methods by which this abstraction problem can be attacked. One begins by determining public goals. In its extreme application, the approach asks individual citizens to work together to form a consensus regarding the direction in which the state or region should move. The goals themselves are only broad expressions of items considered relevant to the general good and they often are evaluated relative to one another.[3] In addition, it may be that some goals act as qualifying statements which prescribe the limits within which others are to operate. For instance, two developmental goals might be: (1) attaining the highest level of noninflationary full employment and (2) the maintenance of environmental quality. Obviously, in some cases the achievement of the first may come at the expense of the second such as through industrial pollution of rivers or a less than optimum rate of resource use. Hence, the employment goal becomes modified to the extent that it functions within the accepted bounds of environmental quality. With some imagination, it is possible to visualize a situation that requires an employment modifier to be attached to the environmental goal.

As the number of goals multiply and the interdependence between them becomes more acute, it is necessary to further refine them in order to evaluate a specific project. The goals must be stated in terms of specific objectives that possess a quantitative or qualitative dimension by which attainment can be judged. The objective becomes the reasonable absolute that specific activities are aimed at reaching. The degree of achievement is measured by the appropriate criteria selected for its ability to indicate the level of attainment. For instance, if happiness is the general goal, then the specific objective might be a given increase or level of leisure time. The criterion might then become the accrued number of leisure hours.

The application of the goal-objective format to transport system planning appears in the evaluation of a given project's ability to contribute to the attainment of the community's objectives. Of course, as noted earlier (see sections 1.4 and 1.6), the comparison of alternative plans requires some way of making relative judgments among varying objectives. While many objectives resist comparability in terms of their measurable dimension, it is important to realize that this form of "top to bottom" evaluation (from goal to item) does direct the decision maker to deal only with those aspects of the supersystem that are relevant to transport planning. If this task of specification can be performed by an appropriate citizen group, then there is little or no confusion over the absolute importance of the items being considered. However, the weakness of this approach is that there is no degree of certainty that this function can be accurately carried out in all cases.

The second methodological approach to the abstraction problem involves the listing of *all* possible interaction impacts originating from the transport plan. As each alternative is considered, the variables affected by it are quantified and used as the basis of the comparison. This approach requires that each potential impact must be outlined in operational detail which allows for its inclusion within the model. There are two serious drawbacks to this "bottom to top" (item to goal) method. The first is that it really does not synthesize very much. True, there are nontransport related events that are not considered; however, those factors which do remain constitute a large number of effects, and their analysis may entail considerable financial costs. There is no indication as to the events upon which the investigation should focus. Rather, quantification efforts are spread thinly over all potential impacts. In some cases, impacts which are usually considered to be important may not be relevant. Consequently, efforts are wasted on needless investigation. To correct for this, information is needed that allows for systematic selection from among the theoretically possible impacts.

The second limitation of the approach is that it involves implicit goal setting. Since it is only through the existence of community goals that the relevance of an impact can be justified, by stating that each deserves inclusion within the decision-making framework, the inference is made that it is related to a specific goal. The potential for confusion and misdirection that this holds for transport planning could be quite serious. For instance, if a given road plan eliminates an historic site, the "bottom to top" technique would require a consideration of this impact within the location decision. However, it does not necessarily follow that this or any other historic site is important to the community. Conceivably, it could be that the community does not value the preservation of historic sites. As a result, the elimination of the site is of no concern to area residents.

Such information could mean the difference between acceptance or rejection of a proposed route. If planners, in quantifying the impact, feel it is of significant community value, they may choose an alternative route, which appears to offer relatively fewer harmful effects. However, given the actual value structure as just outlined, which is unknown to the designers, the substitute choice may inflict real harm or be less beneficial with respect to a variable which is truly valuable to the community. The validity of the route selection process rests on the correct inclusion and relative evaluation of individual impacts in the goal-value structure. Unfortunately, this "bottom to top" approach, which in the extreme gives equal weight to all impacts, makes no attempt to determine the existence of the impact-goal relationship if indeed it can be determined with any certainty.

A further complication is introduced when it is recognized that the quantitative effect of a highway on either the community's goals or impact variables can be either harmful or beneficial. Given the initial conditions within which the community operates, it may be that a beneficial effect on a certain item has a level of importance different from that associated with a detrimental effect upon the same item. For instance, a community goal regarding environmental quality might be so stated as to indicate a desire to maintain an already acceptable level of purity. Activities that raised this level by eliminating current sources of pollution would be of far less importance in their impact than an activity which increased the extent of environmental abuse. As a second example, it is conceivable that a highway can have either beneficial or detrimental effects on the number of area residents employed. Again it could be that there exists a lack of symmetry between the importance of the two opposing impacts. Consequently, the investigator not only has to imply through his research that an item is of importance to the community, but he must also determine if the relative concern attached to the item alters as the nature of the impact changes from being beneficial to detrimental.

The resolution of the goal-impact orientation dilemma is an essential aspect of any highway investment model. The choice of a perspective will certainly influence both the direction of the information gathering process as well as the quality of the analysis. In summarizing the nature of the trade-offs encountered in developing this characteristic, the following comparisons are relevant. With respect to the weaknesses of each, the choice is between an approach where there is a possibility that the participating group may be unable to either articulate or agree upon a unique set of community goals or that the goals may not be translatable into tangible objectives, and an approach that may overextend the limited set of investigative resources by considering all effects equally. On the positive side the comparison is between a technique that focuses the direction of research on the most important areas, and a technique that is based upon fact, not philosophy, and consequently requires only an evaluation of what is rather than a consensus upon what should be.

2.4 The Geographic Perspective of the Model

This characteristic results from the fact that a highway is an immobile facility which occupies a unique location. Consequently, two demands are placed upon a decision-making model. First, the types of impacts experienced and their relative intensity are in part a reflection of the area in which the facility is located. As roads are built in regions with varying economic bases, social backgrounds, population densities, and transportation requirements, identical physical structures can be anticipated to generate different impact patterns. To be fully effective a model must be sensitive to this form of locational variation.

A second demand focuses upon the distribution of impacts. Since travelers need not live next to a roadway in order to use it, this can lead to the appearance of special factors as the key distinction between those who draw net benefits from the highway and those who, on balance, are adversely affected by it. It is not unusual to find that the opponents of a corridor possess the greatest degree of geographic proximity while the advocates are chiefly represented by projected desire lines of travel emanating from outside the immediate area. To account for this location based divergence between the service and proximity effects of a roadway, the analytical technique must alert the decision maker to both the identity of those affected as well as the magnitude of the impact.

The view presented in this section is that in order to meet these demands, the model itself has to assume an explicit geographic perspective. It is offered that this type of perspective is an inherent attribute of all transport models whether expressed or implied, and it is necessary as an organizational device. Since two independent requirements exist, no single perspective may be capable of simultaneously maximizing the amount of analytical effort directed toward each. For example, an approach that fully identifies the totality of impacts as they affect the characteristics of an area may contain a bias in terms of its distributional analysis. Considering the possibility of this inverse interdependence, the objective of the following review is to evaluate the perspectives with respect to their degree of identificational and distributional success.

In general, three perspectives can be distinguished that focus upon the problems under discussion. The first is a statewide view which regards a highway as a means of facilitating the flow of people or goods, both within and through the state. This perspective also recognizes that historically the provision of this form of transport service has generated a degree of economic progress and development for the region being traversed.

In some cases it may appear that this broad view is "forced" upon the planners because they feel unable to avoid what they interpret as the realities of the traffic flow situation. For instance, where the state exists as part of a megalopolis and provides corridor services between major urban concentrations, plans may be predicated upon the assumption that the demand for a facility is at least regional and quite possibly interstate in nature. Consequently, a noticeable volume of traffic can be counted upon to exist even in the absence of a building plan that attempts to provide for it. In response to this supposedly predetermined demand the priorities of the highway planner are turned toward the provision of safe, efficient, and fast transportation through the area while minimizing the inevitable detrimental impacts upon segments of the community.

The chief difficulty inherent within this broad perspective is that while it attempts to locate a route that establishes a balance between user and nonuser effects, the result is often incorrectly weighted in favor of the numerically superior group of potential users. This occurs because the service priority is extended to the point where the corridor location effort is reduced to a process of selecting a site which generates net benefits for the largest number of people. It establishes an implicit ratio that, in the extreme, equates one person harmed with one person benefited, regardless of the interpersonal or relative degree of impact. Consequently, if it can be shown that more people gain from the service impact than are harmed by the proximity impact of a roadway, then the objections of a few are overridden in favor of satisfying the demands of the automotive many.

Clearly, in order to correct this problem it is necessary to incorporate additional information into the model which allows for a more personalized point of view of the impacts of the project. To accomplish this a different geographic perspective is called for. One possibility is the development of a regional approach that involves a parceling of the state into fairly homogeneous districts which reflect prominent area characteristics. The relationship between these characteristics and the roadway then becomes the means by which the value of the facility is assumed. Where conditions permit the characteristics may be based upon the differences in life styles found within the state. For example, in applying the model to a rural as opposed to an urban region, the residents may hold different attitudes regarding road use, location, and nontransport importance.

In urban areas the major user concerns are usually directed toward servicing commuter trips and the distributional needs of business firms. On the nonuser side the emphasis might focus upon the interaction of the facility and the social stability of the area or the financial position of both individuals and the local government. The procedure of the regional approach is to estimate and assign both the service and proximity effects within the area in order to develop a view of the localized net impact. This result can then be compared to the anticipated impacts upon persons outside of the region to show the degree of trade-off between regional and aggregate interests. Ideally this will serve to reveal potential areas of conflict between the two groups.

In rural areas several alternatives are possible regarding residential attitudes towards corridor location. At first glance the area may appear to be agricultural; however, changes in the underlying economic conditions may no longer make such an activity profitable as a primary source of income generation. As a result, the population may regard a highway as a stimulant opening the area to new industry and encouraging residents to commute to urban centers for jobs which were previously inaccessible. In many instances this is the hypothesis held by road planners and it dominated their thinking concerning rural road location planning. Consequently they place considerable weight upon potential proximity benefits as an additional factor offsetting the harmful effects of the corridor.

On the other hand, it may be that the primary values held by rural inhabitants are not economic ones. Rather they may have deliberately chosen to live in a more quiet atmosphere because of the way of life it offers. Environmental purity, informal social relationships, and a lower level of personal competition are all legitimate values that often find their expression in rural areas. The intrusion by a highway with its accompanying developmental impacts may not be viewed as a beneficial force by these people. Instead, they might see the facility as introducing changes in the environment, social structure or land-use patterns that they have already chosen to foresake. As a result, a highway, no matter how well intentioned, would not receive a favorable reception.

By focusing attention on the regional character of the roadway, there is a greater possibility that the planner can develop an indication of the nature and strength of individual attitudes. Also, in performing the initial evaluation at what is still a fairly aggregate level, the process possesses sufficient ability to neutralize any purely parochial interests that might raise pressures in the direction of a basing a transportation decision on local issues. Clearly the likelihood exists that there is an inverse relationship between the extent of local emphasis and the potential for rendering decisions which might neglect the value of the services rendered outside of the area. Geographic bias regardless of the direction in which it is turned can only distort the degree of efficiency with which transportation resources are allocated.

In some instances, where the project is large and its proximity impacts affect a significant number of people in diverse ways, it may be desirable to require that the model focus upon the smallest subdivision—the "community." In cases where several distinct communication links or sociocultural groupings exist within the same township, it is often useful to describe the area in terms of these characteristics and use their geographical aspect as the basis for judging impacts of the project.

The advantage of this approach is that it permits an investigation of locational effects on the most individualistic basis where the ultimate proximity impact resides. This allows for a detailed review of the equity problem which generally occurs whenever families are moved or residential neighborhoods traversed. This ability to identify personal effects facilitates the consideration of some adjustment for them in the planning process. In pointing the model in this direction, planners are not only asked to determine the absolute effect of a project, but also to classify and to consider the composition of the broad aggregates. Who is affected becomes an element to be considered in route location and planning along with the "how" and "how much."

The limitations of this approach appear mainly in the ability of the planner to identify and work within the geographical boundaries of the "community." First, by utilizing nonphysical characteristics as a guide in establishing geographical boundaries for the "community," an investigator may encounter problems created by administration divisions as he attempts to implement the evaluation process. While a single town may possess several distinct "communities," there is the possibility that the community can extend across town borders.

Second, after the physical "community" has been defined, the planner may be unable to identify the spokesman or leaders through which he must work. In many instances urban "communities" lack any formal structure that permits direct contact with area residents. As a result, investigators could be working through self-appointed leaders who fail to truly represent the feelings of the population accurately. The problems which might arise as a result of the potential misinformation could be quite severe.

In summarizing the nature of the problem raised by the assumption of a geographic perspective, it can be stated that it involves the search for a formula which will provide for the socially optimal integration of the broadly dispersed service impacts and the localized proximity effects of a roadway. Clearly the way in which this problem is resolved will have a noticeable influence upon the direction that transport corridor decisions will take. Apparently it is precisely this conflict that lies at the heart of various current objections to the location of segments of Interstate Highway System.[4] User requirements of road travel are aligned against local citizens who not only question the specific location of a given corridor but they also raise doubts concerning the value of road networks as a means of solving the transportation needs of the nation. Without a geographically oriented analytical framework the degree of opposition is obscured and a quantitatively based settlement is difficult.

Previous evaluative techniques have ignored an explicit discussion of their geographic perspective, and as a result they have implicitly resolved the issue in favor of one of the alternatives.[d] Hiding the issue, however, does not make it any less important nor does it reveal its role in what initially appear to be substantially different problems.[e] An evaluative model which is to be applied as a

[d]General equilibrium or goal oriented models are typically based upon a state-wide or larger perspective while socio-economic indices are usually very localized.

[e]In addition to the corridor problems associated with the Interstate Highway program, the controversies over the Cross Florida Barge Canal and the Alaskan oil pipeline each contains a geographic perspective that distinguishes between aggregate and localized interest groups.

tool in practical decision making, must expose this difficulty to facilitate the development of an optimal solution.

2.5 Transportation Investments and Property Rights

It is apparent from the discussion in the previous section that different population segments can be affected in varying proportions by a given project. Consequently, apart from a consideration of the balance between service and proximity effects, the model must possess an additional attribute if it is to reflect the actual distribution of corridor impacts. The presentation of this attribute necessitates the investigation of two concepts regarding the evaluation of road alternatives. The first reflects the fact that the selection of a corridor involves the transfer of property rights, while the second indicates that all claims on "property," when broadly defined, are not of equal strength and therefore do not require the same type of attention in the decision-making process.

The term *property right* is used in this context to signify that the holder of such a right has the privilege of experiencing and disposing of the item in question. Consequently, the right embodies the "value" of the good or service over which it exercises control. Typically the corridor development decision requires the transfer of some forms of property rights under the legal process of eminent domain which provides "fair market compensation" for the reassignment of the use privilege. When these titled rights are infringed upon either in part or in total without compensation, some of the wealth embodied within them has been transferred implicitly between the parties involved.

In past highway cases controversy has usually surrounded the size of the compensation paid for the expropriation of a physical right. This reflects the differences in value that the parties attach to a specific right. However, it is becoming increasingly evident that additional controversy can also be found regarding the scope of the compensation that accompanies the physical transfer of titled property.[5] For example, the sale of a home involves moving expenses, the costs of seeking a new residence and the possibility, due to market imperfections of paying a premium for a new home of equal quality. The compensation price paid for the old home does not usually cover these costs in full, and since they are actual burdens resulting solely from the location decision their impact should be part of the evaluation process.

This legal compensation for titled property can be regarded as only the visible or first level of the property right concept. A second level, often referred to as amenity rights, also requires attention. Attached to the ownership of physical property is the quality of life that its location affords.[6] The precorridor status quo yields an established level of quiet, a degree of privacy and security, or a measure of air purity. Since the enjoyment of this environment requires the ownership of the property, or at least some right to its use, the value of the status quo with respect to quality is reflected within the market value of the property right.

In cases where the location decision affects these amenities, there is a corresponding change in the value of the right. For example, if a corridor passes near but does not physically infringe upon a residential area resulting in exhaust fumes, traffic noise, or increased vehicular flow on feeder routes, this will cause an alteration in the value of the property in its present use. If the market place assesses the net impact of these changes as detrimental, then the current value of the property right whether held or transferred will be diminished. As a direct impact of the corridor location activity, this loss of value rightfully belongs within the evaluation framework. As Mishan argues, since amenities are in themselves scarce goods, their application in one use as opposed to another involves the economic problem of resource allocation.[7] Consequently, the transfer of resources as reflected in the amenity right must be undertaken at a price which indicates their value to society. It is this "economic view" which has to be considered within the investment decision. As of the moment, amenities lack legal status, therefore they are typically excluded from review by the evaluation process.

The final extension of the property right concept involves what might be referred to as social claims. To distinguish these from the two previous categories, they are defined as the rights to the institutions (either formal or informal organizational structures), habits, patterns of customary behavior, or personal contacts which facilitate and are interdependent with daily activities. Examples of social claims are religious or civic affiliations, individual lines of communication, business clientele, service availability and educational or employment opportunities. They comprise the social status quo in the same way that amenity rights represent the quality of the physical environment. In some instances such as access to quality schools or the location in a community which facilitates the development of desirable social and economic characteristics, the worth of these claims can be reflected in the value of titled property. However, in most cases social claims represent personal values apart from physical property, and the intrusion of a roadway can alter them without afflicting property values directly.

Unlike infringements upon amenity rights which predominantly affect those who remain in the area, the disruption of social claims affects both the physically displaced and those remaining. For instance, if a corridor necessitates the removal of residential facilities, then not only is the physical housing destroyed but also the social contacts are disturbed to the extent that the former inhabitants may be scattered outside of the present community. That these contacts are valuable is evidenced by the following facts. First, there is a cost in time and effort if not actual dollars involved in their development. This expenditure is forfeited if these contacts can no longer be continued. Second, if the contacts were maintained and exercised before the intrusion of the highway then this indicates that the development costs were fruitfully spent and the resulting associations possess a degree of worth to those affected. Finally, even in the absence of periodic exercise of established contacts, their existence enriches the range of social choice. Their value lies in that they act as alternative opportunities to assist the holder of the claim in the event that he chooses not to or can not make the primary contact.

The disruption of social claims is also prevalent in cases where corridor development forces the relocation of patrons of a business while at the same time avoiding physical encroachment upon the firm itself. Without the benefit of an established clientele, the future of the business is jeopardized and its market value is correspondingly reduced. Current decision-making techniques neglect to consider and account for this type of cost in the evaluation process, resulting in an underassessment of the degree of social damage inflicted by the decision to build.

Apart from the evidence that the worth of these social claims necessitates their inclusion within the corridor review, a second argument assists in establishing the validity of this point. Normally, events which destroy existing social claims are evolutionary in nature. They occur over an extended period of time which allows for those affected to take compensating actions. Such gradual change can be regarded as the usual course of a dynamic society to which the inhabitants have learned to adapt. On the other hand, transportation related change is discontinuous and its magnitude usually exceeds the smoothing capabilities of the existing adjustment mechanisms. The size of the change when combined with the limited time span within which it occurs strains the absorptive capacity of the individual. In addition both the existence of the event and the distribution of its attendant impacts are under direct human control. Consequently, to avoid the inclusion of confidential social claims biases the evaluation process in favor of the decision to build.

One way in which to affect the inclusion of each level of property rights within the review technique is to classify those impacted according to their relationship with the effects of the highway. Three general groupings can be developed. The first identifies the impacted persons as individuals indicating that the rights which they hold are directly affected by the project in a highly personal or unique way. This classification can be subdivided according to the role played by the individual at the time he is affected. Readily recognizable subdivisions would include persons residing within the impact area and business firms which function in the community. Within this classification the investigative effort is directed towards individual identification of impacted persons and the quantitative development of the beneficial or detrimental effects of the corridor upon their rights.

In many ways, the construction of a framework oriented in the direction of an individualistic evaluation of project effects lends itself to adaption within the already existing framework of cost-benefit analysis (see section 1.1). Often the individual directed effects can be quantified in the market and, as a result, possess the potential to be aggregated with the existing calculations of user effects. The principal innovation on the nonuser side would be the need to specifically identify the rights and individuals affected.

The second classification of impact relationships requires the identification of the holders of infringed property rights as members of a group within the community. Such group impacts arise in that the rights are affected indirectly through a person's association with the group rather than as an individual performing a unique function. The group itself as a union of personal interests becomes the object of impacts either through its existence or purpose.

To illustrate this form of property right impact, it may be that a project will affect religious or ethnic groupings, historic or civic organization and specific age or occupational categories which possess common elements. In some cases these groups are not formally organized, but are identifiable because of the characteristics that serve as the focal point of the projects impact. Frequently, the impact upon these groups escapes proper identification and quantification in the usual sense of cost-benefit analysis. This is due to both the loose nature of the associations and the previously revealed fact that at best the market recognizes only the infringement upon titled rights and some amenity rights. By initiating the corridor evaluation of group impacts from the tripartite property right perspective, these limitations are overcome. The primary review effort is aimed at directly assessing the alterations in group rights which have occurred as the result of the physical presence of the roadway rather than relying upon a secondary or indirect market signal to identify the nature and scope of the change. Once discovered, these infringements are then integrated into the decision-making process.

The third and final classification considers those impacts that are felt by the community as a whole. In a manner similar to the previous grouping, the impacts are relevant because the person's rights originate from his association with an organization (i.e., a special group called the community) and, as a result, he indirectly experiences the effect. It is this tie between the individual and the community which brings about the personal effect and not a unique role played by him.

To demonstrate the application of this concept, there are certain factors or aspects of pollution–aesthetics, project design, safety, and social interaction–which have the potential of affecting the community as a body. The rights to the factors involved are held in common so that the exercise or enjoyment of the right by one does not preclude its use by another. With respect to the infringement of these rights although each member of this special group may not in actuality be affected to the same degree, the key distinguishing characteristic is that the probability of impact incidence is the same for all. Consequently, in accounting for the specific impact upon an individual, the potential implications of it for the community are implicitly recognized.

In summarizing this discussion, the following points require reemphasis. First, the choice of a corridor introduces man-made changes within the existing structure of the community. Such changes either enhance the value of an affected party's position or inflict a cost by infringing unfavorably upon it. The basic premise of this review is that each person holds the right to the circumstances which surround his status quo. Consequently, at the least he should be free of externally controlled alterations which ignore the extent of the change they generate. Second, by focusing upon changes in the status quo through the emphasis upon the transfer of broadly defined property right, the limitations of the market evaluation technique can be avoided. This result is the immediate consequence of the direct appeal to evaluation of physical implications rather than the indirect detection of changes as transmitted through the market. Finally, one practical way of organizing the review of property right transfers is to arrange the impacted parties according to their relationship with the corridor at the time of transfer. Not only does this structuring facilitate the identification of these people, but in addition it conforms to reality by allowing for the fact that not all right transfers require the same type of attention by the decision makers.

2.6 Time As an Element in the Evaluation Process

With respect to a large scale social investment, it can be anticipated that the occurrence of impacts emanating from it do not necessarily happen at a single moment in time. Rather, the consequences of an action possess repercussions that are felt to varying degrees over a period of time. Impacts which occur in this manner can be characterized as being similar to the ripples in a lake caused by the tossing of a stone. It is as incorrect to say that the impact of the stone on the water can be measured by gauging only the size of the first ripple as it is to imply that the relevant nonuser effects are those which occur in the first instance. The repeated occurrence of ripples is a direct parallel to the continuing effects of a highway project on both the physical and social aspects of the community.

An example of both the occurrence and potential importance of this time factor for transport planning appears where families are displaced through the right-of-way acquisition process. If these families possess a low degree of economic or social stability, the resulting disorganization over time may require a continuing effort on the part of highway administrators and other public officials to integrate the displaced persons into a new community setting. No valid measurement of the impact on this group can consider solely the effects during the period of construction. Rather, the real income lost through wealth transfers and the social disorientation caused by relocation are time-related events which must be considered in the long run if the total community impact of a highway investment is to be seen. If such impacts are not considered, they may intensify and grow over time as relocation problems are combined with personal hardships to create undesirable community consequences. Furthermore, unresolved personal injustices may lead individuals to band together in community groups whose sole purpose is the obstruction of future governmental activity in the area. Obviously, dissatisfaction of this type, if it is allowed to appear, can make it difficult to initiate corrective programs at a latter date.

On the beneficial side, several positive impacts of a facility may only manifest themselves in the long run. Obviously, user benefits are continuous into the future, as private and commercial vehicles enjoy the use of the facility throughout its service life. In many cases the construction of a network is justified on the basis that its capacity anticipates the growth in travel demand over some recognizable period of time in the future. As a result, while the immediate beneficial effects are small, in the long run it may be possible to more than offset any harmful impacts.

Other positive community impacts that possess a time-oriented nature would include impacts on the rate of industrial development, the number of persons employed, or the extent of community health and safety. The quantification of each of these factors requires the evaluation of both their immediate as well as their long-run impacts in a way that will permit their proper relative comparison to the detrimental impacts.

Considering the problems that are normally encountered in selecting a methodology for equating present and future benefits and costs, it may be advisable to reduce the margin for error with respect to these factors by engaging in two precautionary actions. The first would be to provide some form of sensitivity analysis which shows the effect of changes in key assumptions and physical impact estimates on the evaluative conclusion. This topic has already been examined in the section dealing with the systems approach to economic evaluation (see section 1.6). The other alternative (which may be used in combination with the first) would be to keep the immediate effects, both beneficial and harmful, separate from those other impacts that it is felt will occur in the future. Obviously, other impacts must be considered in the aggregate when the final decision is made; however, given the greater risk associated with estimates of future as opposed to current impacts, it might prove wise to see what percentage of the total on each side of the ledger is accounted for by the different time-categories of impacts. The construction of a facility with sizable immediate detrimental impacts may owe its justification to projections of estimated nonuser benefits. Given the element of risk involved in making and accounting for these projections, it may be useful for the decision maker to have this information at his disposal when the final judgment is made on the case.

The recognition and integration of this time dimension within the review process presents two problems to the decision maker. The first is the development of some valid estimate of the size of the future impacts. This entails the introduction of one or more assumptions regarding the way in which public officials, including highway officials, react to the situation. As indicated above, if future harmful impacts are ignored, they may either gradually disappear or may fester and grow in magnitude over time. On the other hand, if corrective action is initiated at the inception of the problem and continually applied, then it can be expected that the impacts will decline over time at a rate which will ensure their eventual demise.

Given the magnitude of the estimate, the second problem involves the decision maker in attempting to relate the value of events which are distant in time to those that are more immediate. The technique of cost-benefit analysis outlined in section 1.1 provides one approach to the integration problem. By quantifying impacts in the market, it develops an estimate of the present value of future events by discounting them according to the appropriate rate of interest. Obviously, this solution requires some assumption regarding the usefulness of the market as a measure of value. In eschewing the discount process other evaluative approaches give equal weight to each impact regardless of its timing. They simply require a determination of whether the project will eventually achieve the desired objective at an acceptable level of cost. Implicitly these techniques are assuming that the time horizon for the community effects is short enough so that any differences in timing can be safely ignored. In some cases this may be warranted however, if the exceptions only reveal themselves after the fact, then approaches which utilize this assumption may generate a set of social and economic forces that could be very difficult to control once they have started. So at the very least, some estimate of the risk of this possibility should be present within the analysis.

In summarizing the importance of this attribute, it is clear that it directs attention to determining *when* an event occurs in addition to the conventional assessment as to whether it appears at all. It is precisely this problem of impact timing for both user and nonuser effects which renders many transport investment projects objectionable to those adversely affected. They see not only present disruption but also a potentially disorganized, uncertain future which must be faced without any assistance. The recognition of the continued nature of all phases of transport related impacts would go a long way toward correcting the problem.

2.7 Summary and Conclusions

The attention of this chapter has been directed towards the analysis of various characteristics associated with transport investments. These characteristics are inherent within the nature of the investment activity and hence warrant analysis by the decision-making process. The reason for this detailed review is twofold: First to outline their nature and second to emphasize that frequently either the existence or scope of a characteristic is directly ignored by the current set of evaluative techniques. Unfortunately, the intrinsic nature of the characteristics necessitates the implicit resolution of the issues that they raise without an open discussion or recognition of the potential ramifications of the decision. The consequence of this omission is the development of a bias toward the underreporting of the full range of impacts. Since the incidence of this underreporting is more often borne by those who are impacted in the nonuser capacity, the bias tilts the ultimate decision in favor of the alternative that most suits the demands for nonuser transportation.

To correct for this bias, the characteristics must be explicitly developed as dimensions of the evaluative technique. Such development would integrate them as a central aspect of the review process in a manner that parallels their strategic function within the investment activity per se. The integration will have the effect of emphasizing the importance of the characteristics to more fully reveal the range of impacts which result from the transportation investment decision.

3 The Evaluation Framework: Form and Operation

3.1 Introduction

Utilizing the format presented in chapter 2 as a starting point, the next step involves a description of the structural nature of the evaluative framework and the role that it is to play within the decision-making process. The emphasis of the preceding chapter has been on the development of a set of characteristics which reinforced, in the mind of the reader, both the dominant aspects of the project under investigation and the general characteristics of the resulting analysis. In broad terms these characteristics can be listed as: (1) the presentation and evaluation of the analytical dimensions that serve as relevant abstractions from the supersystem, and (2) the identification and quantification of the transport-generated interactions among and within parts of the supersystem over time. Stated somewhat differently, the items discussed in the background material center on the *flow* of activity that accompanies any change in the transport system and which must be considered rationally within the decision-making process.

To evaluate this horizontal *activity flow*, the investigator must possess a technical tool that allows him to *abstract* the relevant information from the wide spectrum of available data. This abstraction process then enables him both to *identify* and to attempt to *quantify* those changes in the current and projected course of events in the supersystem that can be attributed to alterations in the existing transport network. As noted in the introduction to chapter 2, the only limitation placed on the evaluative process is that it possess analytical and predictive capabilities.

On the surface it would appear that the conceptual background of the model as outlined in chapter 2 provides a neat and concise description of both the duties that the framework must perform and the field within which it is permitted to function. However, upon further examination, it becomes apparent that the *activity flow* is itself only part of the larger *decision-making flow*. Certain elements of this second flow, when viewed with respect to the point at which the evaluative tool itself is introduced, can have a significant constraining effect on the results tabulated by the framework. In reality, if decisions actually are made over time on an incremental basis rather than at an instant in time when all the information has been gathered, then the actions taken early in the vertical flow can act as indirect or unseen limitations on the operation of the evaluative framework by constraining either the type or number of alternatives considered.[1] It appears entirely possible, and indeed probable, that the extent of flexibility at any stage in decision-making stands in direct relation to the choices made at the preceding phase of inquiry.

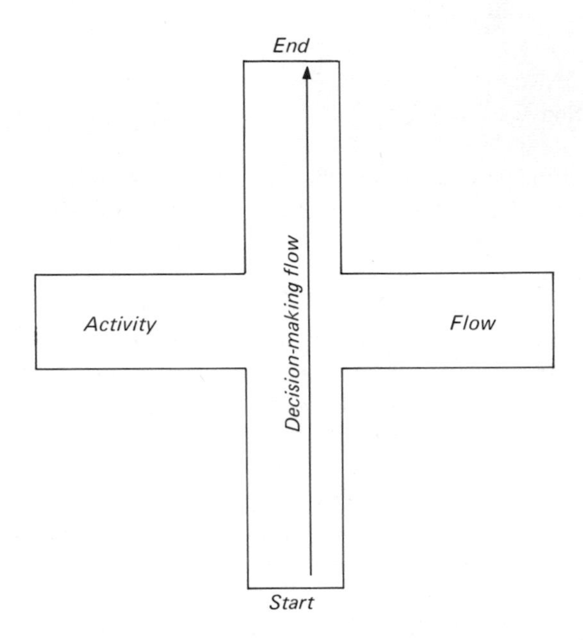

Figure 3-1. *The Combined Horizontal and Vertical Flows of Activity and Decision Making.*

If the above is true, and there is no *a priori* reason to believe that it is not, then the stage at which the framework with its analytical approach and data requirements is introduced becomes a crucial factor in determining both the nature and the quality of the answer that it will yield. As an extreme example of this point, assume that in the planning phase of route location, where the actual development of viable alternatives first takes place, the only factors considered relevant for the proposing of different routes ary the various aspects of traffic service. As a result, the range of corridor alternatives selected and available to the investigator for the evaluation of nonuser impacts is correspondingly limited. Obviously, treating the consideration of the community effects of route location as a residual to be evaluated only after the user aspects have guided and shaped the development of alternative corridors creates severe problems. Nonuser effects enter into the decision-making flow only as a suboptimization where the "best" alternative (with respect to nonuser impact) is selected from a predetermined (i.e., constrained) set of choices. Consequently, there is no guarantee that the chosen corridor represents the maximization of net gain to be achieved on the margin of highway investment. A true maximization process can be introduced only if the user and nonuser effects are considered simultaneously from the beginning.

In conjunction with this point, it is interesting to note that the possibility of maximizing marginal net gain does not necessarily improve as the investigator moves the examination of the horizontal activity flow closer to the origin of the vertical decision-making flow.[2] Rather the partial elimination of constraints introduced by the decisions of previous periods does not introduce total flexibility into impact evaluation; and such flexibility is a necessary requirement of maximization. It may well be that a marginal increase in flexibility will result in a marginal improvement in the quality of the decision; however, there is no substantive way of demonstrating this. The only way to ensure the possibility of maximization is to include the consideration of nonuser effects in their proper perspective from the very beginning.

It is evident, given the previous example, that the investigative tool which will be developed in this chapter is only part of a much larger corridor location picture. The constraints placed upon the usefulness of the tool increase with both the length of time between the origin of the decision-making flow and the application of the tool and the number of intervening stages. Consequently, the nature and reliability of the results are open to qualification to the extent that direct or implied constraints exist. As the evaluative framework is presented in this section, no further extensive reference will be made to the interdependence of the flows. The discussion offered here will serve to indicate a recognition of the point. However, as the structural aspects of the framework are described, their relevance to this question of interdependence will be noted where appropriate.

3.2 Adoption of a Matrix Approach as a Methodological Format

The choice of a methodological approach focuses on one or more of three broad possibilities: (1) reliance upon a market technique, (2) application of a matrix approach, and (3) construction of a functioning structural model. Given the objective of this presentation and the informational needs of transportation decision makers, it was felt that the most efficient method for obtaining the desired result was through the development of a matrix approach. As envisioned, the matrix includes two useful aspects of the other alternatives. First, through an adaptation unique to this analysis, a more extensive use of the market is attempted than has been advocated by previous writers. Secondly, an in-depth, but less formal knowledge regarding the structural relationships between transport and other variables is still required if the data needs of the matrix are to be fully met. Given the blending of approaches, the matrix format offered the least cost method of obtaining the fixed informational requirement.

Other key features also tend to make the matrix approach relatively attractive. First, in terms of operation, the data requirements are separable and their nature is almost totally identifiable from the outset. As a result, the investigator can focus his attention on manageable "bite size" pieces rather than being forced to manipulate all of the elements simultaneously. This segmentation aspect greatly increases the simplicity of operation.

Second, a matrix permits sufficient flexibility to handle the problem of the "geographic subdivision" as discussed in section 2.4. As the initial conditions or objectives of a study alter, a matrix is a more feasible avenue by which to introduce modifications into the evaluation procedure. Such flexibility manifests itself both in the ability to alter the components of the impact frame of reference and in the organizational structure that guides the human subdivision. In other words, it is easier to tailor the approach in marginal ways to fit the required tasks. Given this flexibility, the operational approach to evaluation becomes a systematic plan directed at information gathering that can be repeatedly applied over a wide range of potential transportation problems. Again, the broad nature of the approach and the added familiarity that comes from multiple applications further simplifies and increases the power of the technique.

Third, the accounting nature of a matrix indicates that it is to be used as a tool in the decision-making flow rather than as a substitute for the decision maker. The purpose of this is obvious, since the task is to provide the investigator with an instrument that will assist him in evaluating a given set of alternatives, so that their individual consequences might become readily apparent. In turn, those responsible for making the ultimate choice can then use this technical information as a formal point of departure for offering a final solution.

There can be no single substitute for this final human step in the decision-making flow. No impersonal instrument, no matter how finely developed, can claim total impact sensitivity to the extent that its dictates should be followed without question. Rather a human evaluator is needed to review and improve upon, if necessary, the purely technical results. The maneuvering room necessary for the introduction of this subjective element is available in a matrix presentation. By not attempting to specify formally the structural nature of the relationships with which it deals, a matrix does not demand a slavish acceptance of the fact that it has either fully captured the essence of a project or that it has totally accounted for all the causal relationships. These misleading implications can, however, be drawn from other approaches (see section 1.5). Consequently, the approach opens the door to and fully recognizes the value of subjective decisions. *Nothing* can substitute for professional experience and the rational appraisal of a proposal.

Such activity can be supplemented in a constructive way, however. And this is the role played by a matrix. The results of a matrix analysis point the decision maker in a specific direction based upon the recorded evidence. This evidence is of a quantitative nature and under certain conditions it can be aggregated up to a finite numerical index. A matrix provides more usable information than a mere check list of activity, and it arranges the data in a way that allows for some analysis of the trade-offs to be encountered. But, the indicated choice can be departed from by the decision maker if, in his judgment, the conclusion warrants it. It does, however, become incumbent upon him to explain the reasons for his deviation and to describe explicitly the external factors that swayed the choice away from the technical conclusion. The benefits from this approach are obvious in that the decision maker must be careful in weighting his decision by elements that are open to the full scrutiny of a public review. An example of the need for a subjective decision might arise in a situation where a matrix evaluation based upon the goals and impacts of a project within a subregion indicated the selection of a route that was less desirable from a regional, state or national point of view. In the event of such an occurrence, the decision maker is forced to choose between the conflicting points of view. While the technical case points in one direction, the existence or distribution of important nonlocal factors can outweigh them. The focus of attention then shifts to the relative comparison of impacts by which the decision maker can hope to justify his actions. To the extent that these subjective elements are brought into the limelight, the judgmental process within the decision-making flow is improved.

A fourth advantage introduced by the matrix approach is its requirement that the bits of information can be obtained validly only by extensive field work on the part of the investigator. This field work provides the evaluative process with a research focus which is essential to the development of empirically useful results. In any given situation the investigator must go into the affected community to determine both the "beneficial" and "detrimental" impacts of each proposal and to ascertain the groups and individuals that are on the receiving end. The use of a matrix, with its lack of preconceived notions regarding the size of effects or the structure of the community, places additional emphasis on this micro-oriented activity. Each proposal is in part unique with respect to its quantifiable community effects and, as a result, its consequences must be viewed anew each time.

While repeated application of the same evaluative technique to similar road proposals can aid in the identification of potential effects, the actual quantification of effects requires an on the spot estimation of the *size* of the impact. In demanding a case approach to evaluation, the matrix technique can lead the investigator to consider the effects of a proposal in ways that he had not initially anticipated. Obviously, the personal contact can greatly improve the quality of the alternatives viewed, the number of effects investigated and the usefulness of the final result. This extensive data collection, while time-consuming, provides an added advantage in that the raw inputs used to fill the informational holes of the matrix are themselves carried along in a form that makes them individually retrievable upon demand. The benefits of this are clear in that the aggregation process does not serve to obscure the elements which played a major role in the decision. The visibility of the separate items improves by necessity the care devoted to their determination, since again the results are open to public review.

The final attraction of the matrix approach concerns an aspect which is somewhat unique to this adaptation. In particular, the information provided by the matrix can at times be used as separable items from the full operation of the matrix. This can be undertaken if the investigator is anxious to apply his knowledge at different levels of generality. After the matrix is formally presented in section 3.6 and its operation detailed in section 3.7, a separate section 4.7 will be devoted to an explanation of the levels of use of the information it provides. For the present, let it suffice to note that the operational nature of the matrix and its usefulness within the decision-making flow is not limited to the point at which the tabulation of impacts is undertaken. Rather, the information provided by the matrix has a much wider role to play in helping to shape the ultimate choice.

In summarizing the advantages of a matrix approach, it was felt that, given the need for a tool which was capable of practical application on a continuing basis with reasonable expenditures in both time and money, the proper direction to follow was toward the matrix technique. As indicated from the outset, the purpose of any evaluative tool is to identify and quantify the effects of changes in the transportation network. Clearly, the matrix approach in general places heavy emphasis upon these two goals.

3.3 An Impact Orientation

As indicated in section 2.3, there was a choice to be made between a top-to-bottom approach based upon the operational definition of accepted community goals and a bottom-to-top approach based upon an enumeration of relevant impacts. As indicated at that time, the questions which were considered central to the choice involved exactly: (1) who states that the goals are "accepted"? and (2) how was it to be determined that the impacts are "relevant"? In resolving this difficulty, it was felt that from an operational standpoint the more manageable problems lay on the side of the impact approach. Consequently, the identification of impacts reduces to a listing of the ways by which the community can be affected through a corridor location decision.

Obviously, such a listing must take care to meet two essential requirements: (1) it must be all-encompassing so that the totality of effects is accounted for, and (2) it must be careful to avoid a double counting of the impacts. If these conditions are successfully met, then the result at this stage is similar to the "Planning Balance Sheet" developed by Lichfield.[3] However, without further modification, it is open to the same criticism raised by Hill regarding the planning balance sheet; namely, that the identified impacts lack any basis in the community, (either judgmental or objective), regarding their actual importance.[4] Without a firm relationship between the listed impacts and the goals of the community, there is no factual way of determining either the absolute or the relative importance of each of the items within the decision-making process.

To correct for this limitation, a weighting of each of the items in terms of their relative importance should be undertaken by members of the community. How the weighting is to take place and which "members" of the community are to perform the function are considerations important enough to require separate sections describing the resolution of the problems involved (see sections 4.1 and 4.2). Furthermore, how the items might be selected and structured are topics that are also given individual attention (sections 3.8-3.10). In these sections, one process of item selection and arrangement is shown in detail.

For the moment, it is sufficient to point out that the existence of a set of relative weights developed through community participation implies that each item receiving a weight deserves recognition, since it has as its foundation an accepted goal or value held by the community. Furthermore, the size of the weight partially governs the importance of the impact within the decision-making flow and, through an indirect process, forces the representatives of the community to structure their aspirations for the future course of events in the area into a hierarchy of values. The advantage of acquiring such information in an operationally useful form is obvious, since it enables the investigator both to estimate the relative importance of each of the items and to introduce them as an element in the aggregation process. By accepting a weighting of the impacts after quantification has occurred, the limitations encountered in the planning balance sheet are overcome. Then the importance of each item can be established.

3.4 Dominant Attitudes Toward the Operation and Definition of the Matrix

In developing an evaluative framework, chapter 2 indicated that there were several technical aspects or questions of methodology which must be explicitly resolved and built into the result. In addition to the items considered there, three accompanying "points of view" must be introduced for justifiable reasons. The first of these concerns the way in which the individual items are to be quantified. The relevant goal to strive for is the development of an accounting technique that permits the aggregation of effects up to an index of total impact. This is the point which dominated the research effort. In reviewing the alternatives to aggregation and the difficulties associated with each, the net result was the opinion that the benefits to be derived from an approach incorporating this characteristic far outweighed the disadvantages that could be anticipated. In particular, the introduction of trade-offs through the aggregation process would allow for the comparison of alternatives on the margin of decision making to the extent that the discontinuities in the construction process permit. This improves the quality of the final decision, since the investigator is able to determine more precisely the meaning of changes in either design or location specifications.

The above need for aggregation is a companion to the second "point of view." In essence, this kindred concept states that all the elements normally encountered in highway corridor location decisions can be viewed relative to one another. No single item, solely by its occurrence, has the power to cause the proper authorities to veto the choice of a specific route. If all effects are relative, then there must exist a set of trade-offs under which decision makers are currently operating.[a] The problem with these trade-offs as they presently exist is that they are usually only implied through the decisions rendered. They are not open to direct review and may not be fully ascertainable even after the verdict is rendered. This is due to the multiplicity of contradictory events that exist and which can potentially occur as a result of location decisions. If these trade-offs can at present be seen only indirectly and through a clouded atmosphere at best, then it may well be that the ratios now in use are either not considering the proper factors, relative to each other, or are considering the proper factors relative to each other, but in the wrong proportion.

Given present conditions, either or both of these possibilities can occur and still have their existence hidden by the final result. This condition arises if the size of the error term approaches zero because the deviations introduced by the above points are normally distributed about a unique or generally acceptable set of correct choices. However, there is no guarantee that this occurs in all cases, especially if the problem of reduced flexibility along the decision-making flow exists (see introductory discussion to this chapter). At an early stage, any decision maker possesses the potential to introduce a systematic bias into the flow that can reduce the chances of obtaining a final result which maximizes the net benefit occurring to the community from the facility. Only by accepting the existence of trade-off ratios and striving to make them explicit can the decision-making process hope to improve upon its results with any certainty. The evidence regarding such trade-offs must be brought out into the open, where it can be examined, discussed, and agreed upon by representatives of both the bureau of highways and the communities involved. In doing this, the structural relationships under which the decision maker operates can be mutually reviewed to the benefit of all.

By stating the above "point of view" the implication should not be drawn that a formal development of trade-off ratios is considered to be a simple task or one that can be accomplished through the application of a moderate amount of professional time and effort. There is no denying the fact that the task is a long

[a]These trade-offs are either fixed or variable depending upon both the time horizon and the magnitude of impact.

and difficult one that may require constant effort and attention in order to be operationally feasible. The issue under discussion here is that the efforts of investigators *should be* directed toward an explicit consideration of these trade-offs in a way which makes them intelligible to those who wish to review the results of the evaluative framework. The level of difficulty associated with the chosen goal *should not be* used as an excuse to deter the efforts made in the direction of this achievement.

The final issue concerns the way in which the two previous "points of view" are to find their fulfillment. The question presented here is, "How can the dual needs for aggregation and the development of trade-offs be best achieved?" The answer to this question requires the acceptance of a medium through which the investigator must work. In resolving this issue there are two basic choices: (1) reliance upon some form of market principle, (2) development of trade-offs through a negotiated political settlement. In reviewing the advantages and disadvantages of each approach, it was felt that the elements of openness and objectivity were crucial to the acceptance of one of the alternatives. Again, these elements were chosen as the direction in which the evaluative framework should be pointed rather than hard and fast goals that must be necessarily reached. The effort in this direction is the key element. After a careful review of the situation, it was apparent that a "market test" of project acceptability best performed the task. Even considering the limitations involved, it was felt that the relevant point of departure for the development of a quantified index of total impact was some form of market or quasi-market. The existence of external or nonuser effects makes the task a difficult one (see section 1.1). However, the construction of a framework where dollar accounting played an important, if not complete role, in determining trade-off ratios was considered most advantageous.

The reasons that governed this acceptance of a market-oriented point of view included the following. First, under given assumptions, the market place offers a test of efficiency in the use of resources. It considered which allocation of inputs out of a continuum of available choices will best provide the goods most wanted by society. Then, given that the production of a specific product is deemed most desirable, it offers a test by which the most efficient production technique can be implemented. Obviously, in the evaluation of transport networks the ability to discriminate in this way between alternatives is essential to the attainment of maximum benefit from the facility.

Second, the reliance, at least in part, upon dollars provides a uniform accounting system that holds out the opportunity for aggregation. Furthermore, the existence of a tool for measurement gives some concrete notion of the potential impact and a point of departure for debate if various groups differ in their opinions regarding anticipated effects. By moving in the direction of dollar estimation, the matrix forces the parties involved to formalize their opinions in a finite number or range of numbers and to support them with factual statements and explicit assumptions. If nothing more, the emphasis upon rational analysis will itself help improve the decision-making flow.

Finally, the dollar test of project impact provides both the professionals and citizens involved in a specific corridor location problem with a universally understood unit of measure. The familiarity with the unit allows for a more direct avenue of communication between parties and avoids adding further confusion to what can become an emotionally charged issue. Other decision-making type matrices fail to apply a market test, but prefer to rate alternatives on the basis of either their probability of goal achievement or their estimated effect (subjective) rated upon some absolute scale (see section 1.4). Obviously, the outcome depends in part upon the person who performs the subjective rating and his interpretation of the units involved. Since either of these items can be a source of controversy in determining a final choice, the dissatisfaction associated with the decision-making process can be reduced, by making at least one of them more commonly understood.

Before passing on to the next topic in this section, it might be worthwhile to pause for a moment and to outline some of the characteristics associated with the other means of corridor selection, i.e. a political or negotiated settlement. First, the effects of a project would more likely be noted in a less formal or at least nonmarket terms. Instead of attempting to quantify impacts in dollars, the object simply would be to indicate the physical size of a particular effect. After this rough quantification is made, the results are presented in the political arena (either legislative or ad hoc), where the merits of each alternative are debated, and where the harm caused by each is distributed on some "equitable" basis. The result of the process is to highlight the principal areas of contention and to evolve a negotiated set of trade-offs as each of the problems is answered.

The difficulty of this approach lies in the method by which conflict is resolved. In particular, the possibility exists that the result can be clouded by irrelevant external issues, the inability to measure impacts in a meaningful way or the lack of proper representation for all those involved. Each of these limitations centers around the added element of subjectivity introduced by this alternative. External forces can arise where the focus of attention shifts away from the issues at hand and onto the participants engaging in debate. The contest becomes one of opposing personalities rather than an attempt to reach a judgment based solely upon the facts. At other times, an irrational appeal to emotion can unduly sway a decision. The tendency in such cases is to forget the relative nature of events and to regard one or more items as possessing a near absolute power over the decision, since they elicit an emotional rather than a logical response. Further problems can develop where concurrent, but nonrelevant issues are brought into the discussion. Consideration of these or past events can set up a "logrolling" situation where the bargaining process requires trade-offs outside the transport system. Obviously, the the existence of "riders" can lead to the defeat of a proposal or the selection of something less than the optimal corridor.

A second problem, that of giving a proper voice to the individuals and groups affected, is a major one, especially where the outcome of a project is based upon a combination of the *merits* of the case, the *ability* to present it and the extent to which the speaker is considered as a *serious* contender to be dealt with on a proper level. As indicated previously, the merits of a case should be fairly easy to establish if the party can demonstrate that he is affected. The existence of impacts per se warrants their consideration relative to the other effects. However, in making his case known, the agent affected may not be fully cognizant of the extent of his claim or the ways in which the claim can be presented in the most favorable light. In the extreme, he may fear or fail to understand the negotiation process. The correction of this ability limitation is a large task, but one that is essential to the operation of the approach, since in many instances the people suffering the greatest loss are the ones most deficient in this ability.

Corrective action may require an over reaction in this area to actively seek out the people affected, inform them of the anticipated consequences and to provide them with an organizational structure containing professional advisors who are sympathetic to their needs. The obvious difficulty with this is that community activism can get out of hand and generate a set of cases where none really exist or where the impact is less than presented. The problem then degenerates into a consideration of how seriously each claimant should be considered relative to the others.

The question of "seriousness" in open negotiations implicitly requires a knowledge of the extent of power possessed by each of the contesting parties. The overall balance of power normally goes a long way in determining the ultimate trade-off ratios, and the development of a suitable power level is indeed a difficult item to build into any political negotiation. In the market approach an initial measure of seriousness exists in direct proportion to the quantifiable dollar impact. Those to whom the greatest harm is done should have their cases reviewed closely.[b] The common dollar unit allows this "greatest harm" determination to be made. Other less formal approaches do not contain this index. As a result, the seriousness with which a party's case is taken often hangs on their power base or "clout." In particular, their treatment is a function of the extent to which the party can veto the proposal for all. Those possessing strong veto power often receive undue consideration in the negotiations, while others with less power but stronger relative cases are neglected. A system which permits either extreme to flourish will soon become distrusted and eventually fall by the wayside.

Third, in negotiating a settlement given the above problems, a key issue at stake is often an "equitable" distribution of both the beneficial as well as the detrimental impacts. However, without a proper impact measurement unit the effect of a route is obscured to the point that it becomes impossible to deter-

[b]This can only be used as a rough approximation, since it may be that other affected agents with lower absolute levels of impact also deserve serious attention because the relative burden of the impact upon them is greater. An examination of this possibility requires some knowledge of the personal status of those affected.

mine objectively if the distribution is "fair." Too often, as an alternative, those with the power to do so will distribute the effects as "evenly" as possible hoping that this is a reasonable surrogate for "equitable." Obviously, this is not the case as long as parties possess differing capabilities or capacities to bear up under burdens or to enjoy advantages. If the disadvantages are not recognized as such, then their case will not be heard on a basis proportionate to their needs. Again, the question reduces to a consideration of the aspects of power and ability noted earlier.

Finally, a serious drawback associated with a political or negotiated settlement concerns the tendency to reach a mutually agreeable settlement away from the established forum. Such covert actions often leave the impartial viewer as well as one or more of the parties involved in a position of being unsure as to what the basis of the settlement really was, as well as with a feeling of being shortchanged by the result. The fault again lies with the failure to try and evaluate a proposal by means of a standard yardstick. The lack of a decision-making information mechanism leaves the parties without a single frame of reference to develop impact magnitudes and to serve as a real point of departure from which adjustments can be made if necessary. The application of a market approach yields a more stable base for evaluation and requires detailed field work on a micro level that provides the investigator with vital personal contact with the impacted parties. This avenue of direct communication should be an initial source of stimulation for community interest and participation in the decision-making flow. The result of this contact may very well be that the process of quantification if carried out in a truly honest and effective manner will itself become a forum through which the conflicts raised by the project can be resolved.

3.5 Formalization of the Structural Hypotheses

In developing the "points of view" outlined in the previous section, the end result is a set of formal hypotheses that guided the research activity on the theoretical level and which provided the conceptual framework around which the ideas were organized. A further elaboration and expansion of these hypotheses is provided in operational form in section 4.5. In chapter 5, which deals with the field work of the research, a collection of evidence was gathered which would establish the possible validity or reject as unfounded the operational form of hypothesis stated here. The purpose of the present activity is to describe explicitly the following hypotheses which relate to the construction of the matrix:

1. That the community's own attitude toward the development and impact of a transportation network can and should serve as a guide in evaluating the system

2. That the effects of transport planning and implementation stand relative to one another. This relationship generates a set of trade-offs or weights which are used in the decision-making flow
3. That people possess the potential to feel differently regarding the beneficial and the detrimental effects of highways as they might affect the same factor
4. That the investigator can, through past experience or field work, identify both the impacts of a project and the groups or individuals who are affected by it
5. That individual impacts should not be the sole focal point of investigation. There are certain detrimental, as well as beneficial effects that as they occur in combination will tend to act in a multiplicative manner rather than as simple additive elements
6. That given the need for aggregation up to an index of total impact, a market-oriented approach provides one possible means by which this task can be accomplished

These hypotheses, as stated above, were constructed in a way that would allow for their detailed review. They are offered explicitly in support of the operational format of the matrix-market approach. This approach may yield a substantial improvement in both the investigative activity and the decision-making process as they now stand. Through the validation of the above hypotheses, it is hoped that this will be made clear.

3.6 Presentation of the Matrix

To fully understand the presentation and operation of the matrix, it should be stated from the outset that the basis for the technique is a combination of the identification (i.e. listing) of the possible effects of a project, both beneficial and detrimental, and their valuation within society. How this quantification is to take place is the subject of the following section, 3.7. What is being shown here is a sample presentation of the row and column entries and the organizational structure that governs the accounting aspects of the matrix.

As envisioned, the evaluation of a set of corridor alternatives must review four independent areas of impact. The areas are considered in terms of accounts and include: (1) the beneficial effects of the highway which accrue to the users of the facility; (2) the costs of the facility which can be attributed to or affect the users of the facility; (3) the beneficial effects of a facility which are external to the use or users of the facility; and (4) the detrimental effects of a facility which are external to the use or users of the facility.

The items considered in the first two accounts comprise those effects that are normally included in existing benefit-cost calculations. Brief listings of some of these effects are provided in tables 3-1 and 3-2 (see section 1.1 for analysis in detail). It should be noted that these accounting ledgers remain separate from each other. They also make a distinction between those effects that occur in the present as opposed to those which require both an estimate of future magnitude and a discount rate by which the dollar estimates can be made comparable to the impacts realized immediately. As noted earlier, the separation is maintained to emphasize the possibility of greater risk of variation in those items that have a time dimension associated with them. The outcome of this separation should be a closer evaluation of those projects that necessitate a heavy immediate cost, with a majority of the offsetting returns occurring in the distant and somewhat nebulous future. To correct for the greater risk of error, a sensitivity analysis should be performed showing the effect on the results of alternations in one or

Table 3-1
Benefits to Users of a Facility

		Present Year		Future Benefits Discounted to Present	
Item	Weights	Private	Commercial	Private	Commercial
1. Travel time savings	WA				
a) Driver	WA1				
b) Passenger	WA2				
2. Fixed Vehicle savings	WB				
a) Depreciation	WB1				
b) Insurance	WB2				
3. Variable vehicle savings	WC				
a) Oil	WC1				
b) Gas	WC2				
c)Tires	WC3				
4. Comfort and Convenience	WD				
5. Value of accident reduction	WE				
a) Physical property	WE1				
b) Personal injury	WE2				
6. Overhead Savings	WF				
		Subtotal	Subtotal	Subtotal	Subtotal
					Total___

Table 3-2
User Costs of a Facility

Item	Weights	Present	Future	
1. Construction	WG			
2. Maintenance	WH			
3. Administration	WI			
4. Congestion	WJ			
		Subtotal	Subtotal	Total

more of the key structural assumptions or numerical estimates (see section 1.6 for a discussion of sensitivity analysis). On the benefit side, a structural division between private and commercial vehicles is also advocated, as this would provide some indication of the incidence of positive user effects.

Finally, the column entitled "Weights" contains a set of entries that indicate the relative importance of the estimated user benefits and costs of construction. These weights can be assigned either through citizen participation or by professional highway planners, who estimate the importance that each form of user impact should have on the entire decision-making flow. The technical procedure used to develop the set of weights can be an adaptation of the format outlined in chapter 4. The exact method would, in part, be a function of the nature of the group performing the activity. To avoid unnecessary repetition, the examination of the weighting procedure will be postponed until the previously noted sections are reached.

It should be emphasized however, that at this initial stage of the accounting analysis no attempt is made to net benefits and costs to yield a single figure indicating the size of the surplus or the magnitude of the ratio involved. What is being shown here is simply a monetary accounting of each of the anticipated user effects. As noted earlier, one of the primary advantages of a matrix approach is the open presentation of information that can be referred to with ease throughout the ensuing stages of the decision-making flow. By necessity, if the evaluative framework is to function effectively, this accounting stage must be performed with exceptional care.

Considering the extent of the existing literature on the topic of user cost-benefit calculations, it was assumed that sufficient attention had already been devoted to this area. This permitted the direction of effort toward the development of a quasi-market approach to the nonuser effects, which would allow for their aggregation with the user impacts. In the limited number of areas where this assumption does not fully hold, estimates of possible user effects can be obtained through a sensitivity analysis that maps out a range of potential dollar impacts. This range can then serve as a means of introducing the value of the item into the analysis.

The remaining two accounts of the four-pronged evaluative approach are presented in tables 3-3 and 3-4. While they are identical in their organizational format, table 3-3 requires that the effects of the project be considered solely from the beneficial side, while table 3-4 calls for an accounting from the detrimental side. As noted in section 3.3, the matrix focuses upon the identification of a list of potential impacts caused by highway construction. These impact factors appear on the far left-hand side of the matrix. They serve as the row elements and are arranged into separate organizational categories according to one or more dominant characteristics. Each of these entries possesses a weight that indicates its relative importance within the decision-making process. These weights are signified by the set of small (w's) appearing in the second column from the left. The larger (W's) indicate the weights attached to each of the categories containing the items. The purpose of these weights is further explained in section 3.7, while a possible method for their determination is outlined in chapter 4.

The number of impact factors is solely a function of the number of possible effects that a particular project might have. While the number of impact categories is a professional judgment based upon the anticipated extent of organization required. The existence of both a beneficial and detrimental citation for each of the possible impacts indicates a debit and credit accounting format that necessitates the determination of all impacts. At this recording stage of the investigation, there is no attempt to net the beneficial effects of one impact against the harmful effects. For example, if the facility can be cited as the direct cause of the construction of 100 new homes, the market value of this item would be entered in table 3-3 with no attempt at balancing it against the 50 homes taken for the right of way recorded in table 3-4. Finally, it is feasible that the relative weights accorded the beneficial and detrimental effects are different, especially if present conditions in the community are taken as a starting point. Referring back to the previous example, the loss of 50 homes immediately may have a greater relative weight than the gain of 100 homes due to greater accessibility, because elements of quality and availability would enter into the relative evaluation of the housing impact.

The third column considers the first of three entries of impacted agents. Individual impacts consider those effects of a highway that accrue *directly* to agents due to some definable characteristics that outlines their position or role in the region. Several such classifications of characteristics could conceivably be developed with the one most applicable to a given situation being chosen. For example, those individuals directly impacted might be categorized along one of the five following lines: age, income, assets, education, or community function. In developing these categorical divisions, each person is classified according to the chosen characteristics and the impacts upon him are recorded in the appropriate subcolumn. To further demonstrate the approach, the matrix shown in tables 3-3 and 3-4 offers one of the classifications, community function, in operational detail. The subheadings and operational definitions are:

Table 3-3
Matrix of Beneficial Nonuser Effects

Weighted Impact Categories and Subfactors	Impacted Organizational Structure	Impacted Agents																					Adjustment Elements						TOTAL
		Individual Impacts (I)																			Group Impacts (GI)	Community Impacts (GI)	Impact Interaction (II)			Impact Discount (ID)			
		Residents				Business								Institutions															
Categories and Items	Weights	Owners		Renters		R		M		D		S		Pub.		Priv.		Trans.											
		M	NM	M	NM	M	NM	M	NM	M	NM	M	NM	M	NM	M	NM	S	D	P			I	GI	CI	I	GI	CI	
Category I	WI																												
1.	w1																												
2.	w2																												
.	.																												
.	.																												
.	.																												
A	wA																												
Category II	WII																												
1.	w1																												
2.	w2																												
.	.																												
.	.																												
.	.																												
B	wB																												
Category III	WIII																												
1.	w1																												
2.	w2																												
.	.																												
.	.																												
.	.																												
C	wC																												
Category IV	WIV																												
1.	w1																												
2.	w2																												
.	.																												
.	.																												
.	.																												
D	wD																												
Category V	WV																												
1.	w1																												
2.	w2																												
.	.																												
.	.																												
.	.																												
Z	wZ																												
Weighted Subtotal																													

Table 3-4
Matrix of Detrimental Nonuser Effects

Weighted Impact Categories and Subfactors / Impacted Organizational Structure		Impacted Agents																					Adjustment Elements						TOTAL
		Individual Impacts (I)																			Group Impacts (GI)	Community Impacts (GI)	Impact Interaction (II)			Impact Discount (ID)			
		Residents				Business								Institutions															
Categories and Items	Weights	Owners		Renters		R		M		D		S		Pub.		Priv.		Trans.											
		M	NM	M	NM	M	NM	M	NM	M	NM	M	NM	M	NM	M	NM	S	D	P			I	GI	CI	I	GI	CI	
Category I	WI																												
1.	w1																												
2.	w2																												
.	.																												
.	.																												
.	.																												
A	wA																												
Category II	WII																												
1.	w1																												
2.	w2																												
.	.																												
.	.																												
.	.																												
B	wB																												
Category III	WIII																												
1.	w1																												
2.	w2																												
.	.																												
.	.																												
.	.																												
C	wC																												
Category IV	WIV																												
1.	w1																												
2.	w2																												
.	.																												
.	.																												
.	.																												
D	wD																												
Category V	WV																												
1.	w1																												
2.	w2																												
.	.																												
.	.																												
.	.																												
Z	wZ																												
Weighted Subtotal																													

I. Resident–Agents intially domiciled in the affected area.
    A. Owners–Persons having a claim on the place where they reside.
        1. Moved–Those actually displaced.
        2. Not Moved–Those affected, but not displaced from their homes.
    B. Renters–Persons who live in a facility on a contractual basis.
        1. Moved–Those actually displaced.
        2. Not Moved–Those affected, but not displaced from their homes.
II. Businesses–An Agent performing a profit oriented activity with facilities located in the affected area.
    A. Retail–Agents that sell final goods.
        1. Moved–Those actually displaced from their existing facilities.
        2. Not Moved–Those affected, but not forced to relocate their facilities.
    B. Manufacturing–Agents engaged in one or more stages of final goods production with facilities located in the affected area.
        1. Moved–Those actually displaced from their existing facilities.
        2. Not Moved–Those affected, but not displaced from their facilities.
    C. Distribution–Agents engaged in the wholesale, transfer or storage of goods with facilities located in the affected area.
        1. Moved–Those actually displaced from their existing facilities.
        2. Not Moved–Those affected, but not displaced from their existing facilities.
    D. Service–Agents with facilities in the affected area who offer a skill to buyers in exchange for payment.
        1. Moved–Those actually displaced from their existing facilities.
        2. Not Moved–Those affected, but not displaced from their existing facilities.
III. Institutions–Non-profit organizations with facilities located in the affected area.
    A. Public–Community or government owned facilities (hospitals, schools, etc.).
        1. Moved–Those actually displaced from their existing facilities.
        2. Not Moved–Those affected, but not displaced from their existing facilities.
    B. Private–Facilities held by individuals or groups within the community (schools, churches, clubs, etc.).
        1. Moved–Those actually displaced from their existing facilities.
        2. Not Moved–Those affected but not displaced from their existing facilities.
IV. Transients–Agents who are located outside the impact area, but do part of their business or have contracts within the affected community.
    A. Salesmen–Affected agents who enter the community to sell final goods.
    B. Distributors–Affected agents who enter into the community to aid the transfer of intermediate goods.

C. Personal–Individually motivated agents who are affected as a result of their entry into the impact area.
   1. Employees–Affected workers who enter into the area to offer their services.
   2. Travelers–Those who pass through or enter into the community and are affected by the existence of the highway.

All those whose residential or work patterns are in some way directly or personally affected by the construction of the facility can have this impact recorded in the appropriate subdivision. As indicated, three general classifications of individuals with separable roles are considered as operating within the community. The role of resident can easily be distinguished from that of business or transient, with the tabulated impact of each group of individuals providing a useful indication of the types of persons who are being asked to bear the burden or receiving the benefit of a new facility. Obviously, such a detailed classification of impacts according to community function provides an essential profile of one of the distributional aspects of locational impact. Where necessary, this information could be supplemented with a breakdown according to income or assets to further view the distributional effect.

The final classification of those directly affected requires the consideration of the impact of construction on physical institutions within the community. Such impacts can be physical or structural, with the latter occurring when the characteristics of the clientele being served by the institution are altered. Such characteristics could include the age, income or ethnic distribution of the group as well as its numerical size, religious or educational background. Each of these changes will necessitate alterations in the operational format of the institution. Such changes can run from highly beneficial to some which are minor or even those that threaten the very existence of the facility. Obviously, these effects are important to the community and should assume their proper place within the decision-making flow.

The next category of impacted agents considers the effects of the facility on recognized groups within the community. Such groups require consideration as a body since the individual members are affected only to the extent that they participate in the group. In other words, the effects upon individuals are *indirect* through their association with the group. As a result, there is no attempt in recording impacts to separate out individual effects, but rather the group is treated as a single unit. A list of groups requiring attention in the above manner would include:

1. Historical societies
2. Recreation and cultural organizations
3. Ethnic and religious groups
4. Conservation organizations

5. Education related groups
6. Ad hoc committees organized around a specific issue
7. Social, business or fraternal clubs

These groups may be organized on a national, local or regional level with a recognized following in the area. The only important consideration is that they be affected in some observable way by the location of a highway.

The final category of impacted agents considers the effect of a highway from the overall community point of view. In this case the "group" is the entire community with individual members being affected solely through their contact with the community. The effects recorded in this category possess the potential of touching each and every agent in various degrees. Such effects may result in alterations in the life style of community agents, implying a change in either the quality or form of living in the area. Again, the emphasis is not on the individual being directly affected, but rather the potential for impacts shared in common by all. Examples of community-wide impacts would include: health factors, safety effects, impacts on the social structure, and impacts on the physical characteristics of the community. Regarding the construction of any given facility, perhaps only a few of the effects would warrant community-wide accounting. However, information regarding the existence and magnitude of such impacts is essential, since the impact at times provides a source of common bond to unite the community in favor of or in opposition to a specific route. This unification can result in undue pressure for approval or rejection that must be placed in its proper perspective by the investigator.

In relating this format to the discussion of property rights in section 2.5, it is evident that each of the points of view discussed at that time have been built into the matrix. The reason for doing this rests on the assumption that there are distinct levels of human impact that must be accounted for. These levels are not satisfactorily tabulated by forcing all individuals into one of several rigid categories *or* by regarding all impacted agents as separate entities. The intermingling of individual, group, and community impacts recognizes the basic social nature of man and advocates a consideration of effects within this social structure when appropriate. In taking this stand, the evaluative matrix places an emphasis upon the interaction of the social and transport systems.

Two additional columns of information are considered under the heading "Adjustment Elements." These entries take notice of two shortcomings that require special emphasis. The first recognizes the existence of an "Impact Interaction Adjustment" which allows for the tabulation of any multiplicative element which might enter into the evaluation. By focusing upon individual impacts the tendency arises to overlook the compound effect upon the individual if, due to corridor route selection, he is impacted in more than one way. For instance, it can be anticipated that an agent who has both his employment and life style activities distributed by the facility will experience a greater change in his well-being than could be judged by simply noting the occurrence of individual effects. Again, the result is to focus attention upon the fact that those affected are not faceless entities lacking substance as they appear in the matrix. Rather, the impacted are people or aggregations of people and must be recognized as such. By forcing the investigation to keep an accurate account of the identity of those affected and the nature of the impacts, a more realistic estimation of the nonuser effects of the facility can be obtained. Any multiplicative elements that are detected in the accounting process as a result of the investigation of the community impacts, are accounted for by recording them in the "Impact Interaction" column.

The final column separates out those impacts which, because of their occurrence or continuation into the future, require a discount rate to be applied. This rate, as noted earlier, takes into account the fact that events which occur in the future do not deserve the same weight as current events. Consequently, a "discount factor" is used to adjust the future events so that they may be equated with and evaluated in the present. This addition of the time element as a separate column entry is undertaken for two reasons. First, it clearly shows the key role played by the absolute size of the "discount factor." By isolating this item the importance of variations in this number can be further shown. Secondly, the separate column forces the investigator to be continuously aware of the time dimension attached to each of the identified impacts. To properly complete the task assigned to him, the investigator must ask himself if there are any repercussions from each of the impacts that occur only after a period of time has elapsed. If impacts do occur or continue into the future, then they are accounted for in the "Impact Discount" column.

In varying situations it can be anticipated that one or more of the impacts in both the physical and social spheres will possess a time dimension. This is especially true in cases where the facility affects some facet of the life style of part or all of the community. For example, in the physical system the polluting effect of exhaust emissions is a continuing impact that varies increasingly with the use of the facility. To consider this relative to current effects requires the application of a "discount factor."[c] In the other extreme, personal loss of a physical or social nature incurred due to neighborhood disruption can require an extended period of time to be resolved in a satisfactory manner. In fact, failure to recognize this on-going characteristic of the impact may very well lead to its magnifications over time as it combines with other real or imaginery injustices. To avoid this problem and to explicitly include these external effects within the decision-making flow is a goal of the matrix.

Again, the structure of the matrix serves to point out that the emphasis of the evaluation should rest on the insistence that the decision-making process con-

[c]It is considered increasingly rather than proportionately since the volume of emissions may rise faster than the relative increase in the number of users due to delays caused by congestion problems.

sider the effect of the transport system upon real identifiable agents. These community members should, moreover, be viewed within the physical and social context where they are presently residing. The purpose of this approach is to avoid a faceless counting of numbers that relies solely upon the *quantitative* aspects of network impact and substitutes in its place an accounting based upon the *qualitative* nature of the social and physical effects of transport changes.

The final item to be considered in this section involves the breadth of the approach. The various alternatives emphasize the choice between a single highly flexible technique or an approach based upon the development of a number of techniques each directed towards a particular type of project. The matrix, as presented here, conforms to the highly flexible alternative since its two key features, while repeated in each case, require adaptation to particular situations by the investigator. First, the matrix is an accounting framework which forces the investigator to apply the most informative point of view toward the evaluation process. To function properly, the investigator must realize that he is working with a flow process which deals with recognizable agents and, therefore, necessitates extensive and direct field work. This emphasis upon impacted agents makes the investigator aware of the individual variations encountered in each project. Hence, the investigator is provided with a technique to carry out evaluation rather than an insensitive program to be rigidly followed in all cases. This flow concept combined with the single accounting framework does provide the investigator with a unique general format, but it is to be applied to the investigative process as he deems necessary. As a result, flexibility is combined with the operational advantages of a single investigative approach.

The second feature, which also serves as a source of flexibility, is the evaluation of a project based upon identifiable impacts. These impacts obviously vary as the location and design features of the facility are altered. As a result, the investigator must start fresh in each case to identify and attempt to quantify the relevant effects. In doing this the investigator must be sensitive enough to the existing nature of the supersystem to recognize aspects of a project which are unique to a particular situation. Although a check list of common potential effects is advisable, the existence of unique elements requires that they also be included if the evaluation is to be effective. Hence, additional flexibility is encountered in the form of variable row elements.

In summarizing the above discussion, it is quite clear that the nature of the matrix is problem oriented. It requires the investigator to be both fully skilled in the technical aspects of project implementation and informed regarding the initial condition of the affected community. He must be sensitive to the potential current and long-run effects of the facility as well and be able to move with ease within the community to identify the agents impacted. The matrix provides the investigator with a point of departure that organizes the information, helps to analyze it, and allows for a relative comparison of alternatives enabling a choice to be made. However, it emphasizes data collection that can be accomplished only by extensive field work. The quality of the evaluation rests upon this field work and the opportunity for flexibility that it introduces.

3.7 The Operation of the Matrix

The task of this section is to describe the operation of the matrix as it records the "value" of each alternative with respect to the impacted agents. The previous discussion conveyed the idea that the investigator is to identify the relevant impacts of the project and to assign them to various agents within the community according to their incidence. This is achieved through the interaction of the row and column elements of the matrix. The next step is to quantify the magnitude of the impacts in a way that permits aggregation. In section 3.4 it was explained that the adoption of a market approach to the quantification problem is preferred. In choosing this technique the unit of account becomes the dollar, with aggregation occurring through some dollar weighted index of overall impact.

The implementation of the market approach is carried out by the matrix through the application of a variation in the compensation principle of impact determination. In brief, the principle itself asks if the gain accruing to the beneficiaries of a proposed project is sufficient to allow them to compensate, through money payments or other actions, those harmed, and to do this to such an extent that the injured no longer oppose or feel adversely affected by the project. If such compensation can occur, and the former beneficiaries still feel that they are better off under the change, then the project is considered of sufficient absolute social worth to justify its construction. As noted earlier (in section 1.1), this principle forms the heart of the cost-benefit analysis concept. However, it is limited by the following difficulties:

1. Since it returns the injured to their previous position, it is biased in favor of the status quo. There is no attempt to consider whether this particular compensating move is in itself desirable.
2. It implicitly assumes the ability to make interpersonal welfare comparisons by indicating that the value of the compensated item given up by the beneficiaries is of equal importance to those receiving. In most precise language, it assumes that each person possesses the same capacity to enjoy objects possessing utility and that the dollar transferred yield the same amount of utility to all.
3. It is limited in the practical sense since there is no existing mechanism by which such compensation could be arranged and made in a factual example.

In adapting this principle to the area of highway corridor evaluation, the notion of compensation is retained without specifically advocating that payment actually be made or identifying either the source or mechanism for adjustment.

The idea of compensation as applied through the matrix simply states that there are nonuser effects of construction whose impact upon either the physical or social system can be mirrored in the marketplace by some type of compensating adjustment. This adjustment can take the form of correction where harm is involved or the estimation of benefit where gain is encountered. The important point is that wherever an impact can be assigned to an agent within the community, the proper compensating market force should be estimated.

At present, many instances of actual compensation are already taking place. These occur mainly on the cost side where agents in the community receive monetary allotments in recognition of the costs that are encountered by them due to the highway facility. These include payments for land purchases, moving expenses, and differences in housing costs. However, there is a broad area of project impacts that is not presently compensated for and which requires an *imputation* or an estimate of the dollar amount it would take to correct for the condition.[5] This imputed dollar estimate could be obtained by costing the physical effort needed to correct the situation or judging the amount of money required by those impacted to allow them to take appropriate compensating action. Whatever the source of the compensation estimate, it is important that it be included in the evaluation in such a way that a relative comparison with other impacts is permitted. It is not proper to exclude these items from consideration as the existing techniques imply.[6] They must be brought within the total impact picture even if this can be accomplished only through "best estimate" procedures. Including them initially in the decision-making flow in an imperfect manner and striving for future clarification is a more desirable alternative than allowing them to exist at the whim of the decision maker.

The process through which the above compensation is introduced involves the construction of an artificial index of relative impact that estimates the value of an effect to the impacted agents. This artificial process is so named because it extends beyond the normal type of market transactions encountered. In practice, it is composed of two parts: (1) the estimate, through a "quasi-market" technique, of the dollar payment required to compensate for the impact, and (2) the estimation of the relative importance of the impact and the accompanying compensation to those affected (this is indicated by the weights attached to each of the items by "representatives" of the community). The interaction of these two aspects of impact determination allows the investigator to attach a number to each of the effects that indicates the value of the item within the decision-making flow. To demonstrate this more clearly the process is shown by formula I.

Formula I The Weighting Equation

The relative importance of the compensation per dollar of compensation	X	artificially determined market compensation in dollars	=	artificially weighted relative impact

This formula requires the weighting of the artifically determined market compensation expressed in dollars by the relative importance of the compensation per dollar of impact. The result is a number indicating the artificially weighted relative importance of the impact. This number is then entered into the matrix and is aggregated with other weighted impacts similarly derived. The result of the aggregation process is a numerical value that reflects the total nonuser impact of the project. To make the result compatible with the user impacts shown in tables 3-1 and 3-2 simply requires another weighting based upon the relative importance per dollar of impact of user and nonuser effects. Adding the two sets of calculations together, the overall net results are as shown in table 3-5. This information allows the investigator to view the total effect of each alternative and to make comparisons on the margin to enable choice between competing corridors.

Table 3-5
The Aggregation of Both the User and Nonuser Effects

	Beneficial	Detrimental	B-D
User			net
Nonuser			net
	Total Benefit	Total Detriment	Overall Net Effect

The above outline of the overall procedure serves as the basis for a detailed examination of the steps to be followed in the evaluation process. To facilitate a more organized discussion, the entire process can be divided into seven separate operational steps:

1. Identify the totality of impacts (user and nonuser) generated by the selection of a specific corridor alternative
2. Assign the impact to the appropriate agent within the community according to incidence
3. Develop a dollar figure, either actual or imputed, which represents the amount of money required to adjust for the impact
4. Develop a set of weights which indicate the relative importance of each item per dollar of estimated impact
5. Weight each of the estimated compensation values by the relative importance of the item
6. Aggregate up to an index of total impact
7. Evaluate the corridor route alternatives relative to each other

The first two stages, the *Identification Phase* and the *Assignment Phase*, have been described in section 3.6. The added importance attached to each of these operations is shown by the fact that they provide the ground work for the remaining operations. The quality of the evaluation hinges upon the extent to which the investigator conscientiously performs the basic tasks.

The third or *Artificial Market Phase*, already described, can further be developed along the following lines. Assume that the construction of the facility will inflict an identifiably nonuser cost on one or more agents within the community. In particular, the acquisition of land for the right of way necessitates the destruction of an existing public recreational or open space facility within the area. Part of the impact might be assigned to recognizable groups who use the recreational facilities. The remainder of the impact might be allocated to the community as a whole, because without the facility the area may not be as desirable a place to live or work. Obviously, this is an *apparent* real cost to the community that should be included within the evaluation process.

The quantification of this impact is undertaken through an application of the modified compensation principle in the following manner. Since the impact falls directly upon agents other than individuals, there is no consideration of direct monetary payments to separate persons. Rather, the impacted agents are compensated by physically correcting for the loss of land. This physical activity then becomes the item costed within the matrix. For instance, to compensate for the harm, comparable land sites might be purchased within the community and devoted to conservation use. Or the cost of constructing additional facilities that provide similar recreational advantages might be entered into the matrix. The goal is that, by having the highway authorities extend their activity to these functions (or at least assuming that they have this power), an estimate of the compensation cost can be developed.

In cases where the impact is of such a nature that the replacement cost of the item is incalculable or meaningless, an alternative means of estimating the market value of the item can be applied. This alternative rests upon the assumption that the existing state of engineering technology has within its power the ability to develop a substitute design that will avoid interfering with the item in question. For the most part, this procedure would be introduced in cases where irreplaceable historic sites were involved or where irreparable damage would be done to the aesthetic value of the surrounding country side. The market value of either of these two items relative to the corridor selection decision would be at most equal to the cost of circumventing the object entirely by engaging in additional construction. Such construction would be arrived at by passing the impacted object entirely, resulting in an avoidance of the anticipated damage.

The desirability of this alternative cost approach lies in the fact that it eliminates the necessity of directly estimating the true dollar value of an object for which it would be difficult to create even an artifical market. The important point is to estimate the value of the object relative to a specific transportation project. This is achieved by costing the technical and social aspects of avoiding a given harmful effect. Such a cost estimate will provide an upper limit for the value of the object as it is assessed within the decision making flow. This is a true impact cost limit because if the actual market value of the object exceeds the circumvention costs, it would pay to build the by-pass and leave the object intact.

At present, no infallible method exists for guaranteeing that the costing of either of the two compensation approaches will yield precisely the same satisfaction to the agents involved as the object destroyed. However, such exactness is not the point at issue. What is important is the identification, quantification, and aggregation of previously external impacts so that in general the total effect of the facility can be estimated. Once the above goals have been achieved, the investigator can be permitted the luxury of striving for the perfection of the quantification process. As technical competence in the area improves, it may be possible to bracket the true level of compensated satisfaction by providing both high and low estimates of the activity that would have to be undertaken if the physical harm is to be corrected. This is a further application of the sensitivity analysis approach described earlier (see section 1.6). By viewing these estimates within the overall flow, the extent to which they are crucial to the final decision becomes apparent.

Although considerable interest has recently been generated by the existence of community costs due to highway construction, the body of supporting evidence indicates that noticeable external community benefits are also to be gained from highways (see section 1.2). Such benefits extend beyond the normal list of transferred user advantages and result in an improvement of either the social or physical environment in part or all of the area. It is as necessary to formally consider these impacts within the evaluation process as it is to include the anticipated external costs. These benefits are also open to quantification in the artificial market through a two-step process of impact estimation. These steps include the estimate of the dollar cost associated with the undertaking of a substitute effort that would generate a comparable amount of beneficial impact, and an estimate of the net dollar benefit of the impact itself.

Again, an example may help to clarify the operation of the process. Assume that the investigation in the identification and assignment phases leads to the conclusion that the facility has generated a certain number of new jobs in the community. These jobs are permanent, are a net addition to the economy of the region and owe their existence to the construction of the highway. Such employment impacts *can* be of great importance to the community and consequently they deserve inclusion within the evaluative framework. The quantification of their importance would proceed along the following lines. First, an estimate is made of the type of activity which would be needed, apart from the highway construction, to generate a similar number of jobs of like quality. This compensating activity might take the form of either advertising the present advantages of the area or granting tax concessions that result in the attraction of new firms into the area. This activity is then costed in the artificial market and its value entered into the benefit calculation. This entry serves as an indication of the cost saving associated with the particular realized side effect of the facility. As a second benefit entry, the net discounted present value of the added jobs is added under the heading for time adjustments. Such a figure could be derived by estimating the difference in yearly wages between the employment of the human resource in the newly created job as opposed to its alternative use.

By estimating the dollar value attached to each of these activities, the magnitude of the beneficial impact upon jobs and its role within the decision-making flow can be seen. As with the cost side, the removal of this much money from those impacted would apparently return them to their preimpact level of satisfaction. Again, the estimate is a rough one where limits might be introduced or where consultation with members of the community might be needed to determine the imputed dollar value entered into the matrix. However the problem is resolved, the result is that the particular nonuser effect investigated (i.e., increasing the number of jobs) has been quantified and added into the evaluation process. The superiority of this approach over one that simply notes the existence of the impact is obvious. It can be expected that adoption of this technique, and its repeated application over time, will lead to a refinement of the two-step estimating process. This will occur as lower cost methods of achieving the same impact are substituted and as information regarding past choices becomes clear. This information can be used as a check on previous estimates, which can in turn lead to the correction of current imputed values. Through this corrective mechanism the entire evaluation process can be continuously improved and updated once it has been adopted.

The introduction of a fourth or *Weight Development Phase* attempts to solve a serious problem introduced by the reliance upon impact identification. This problem is generated by the fact that it is impossible for the investigator to determine the importance of the identified impact to those affected. This is a severe limitation since it introduces two new complications into the evaluation process: (1) Does the recorded impact have any absolute importance of its own (i.e., should it be considered at all)? and: (2) How important is the impact relative to the others. Obviously, the first question must be resolved before the second can even be considered.

The seriousness of this problem can be seen by viewing it in relation to the two previous examples. On the detrimental side either of the following extremes can exist under certain conditions. It may be that the recreational land destroyed is the only present section of community property devoted to this use. Furthermore, if it is heavily used by many segments of the community, its replacement might be an essential aspect of construction, even when considered relative to other effects of the project. As a result, the imputed value assigned to this item should be made an important element in the cost or detrimental side of the calculation. On the other hand, it may be that the land taken represents only a small part of a substantial amount of available recreation facilities. Moreover, if people fail to use the facility or they find that the substitute use of existing facilities for the one lost, is acceptable, then replacement of the facility should be downgraded within the evaluation process. The need for an indicator of importance arises from the artificial means by which the imputed value is derived, since any appeal to the market as a test of value must answer the following three questions: (1) Is the item important? (2) What is its dollar importance? (3) How important is the item relative to alternative actions? In the quasi-market structure provided by the matrix, the first and last questions can only be answered by the members of the impacted area.

On the benefit side, the same need for an indicator of importance arises. When the impact upon jobs is considered, two extremes again appear. First, the area may be economically depressed with a high rate of unemployment. Any activity that generates an increase in the number of available jobs would be highly valued by area residents, hence the employment-creating impact of the highway should be duly emphasized within the evaluation. On the other hand, the available work force in the area may already be fully employed, with the added jobs only causing a labor shortage and increasing production costs. In this extreme, the benefit to employment noted by the investigator is of little importance and should not be used in its fully quantified state as a justification for the costs incurred. Some adjustment mechanism must be introduced by which the importance of the impact can be determined and evaluated.

The solution to this problem of impact evaluation is achieved by allowing the people in the affected area to view the tabulated list of physical impacts and asking them to weight these results by the relative importance of each item per dollar of estimated impact. In performing this activity, the participants will be indicating if the registered impact has any value within the community, and if it does, how important it is when considered relative to other events. Obviously, the actual development of a set of weights for the identified impacts is a crucial phase of the evaluation. Further attention is devoted to this topic in part 2 of the book, where one possible weighting methodology is developed and applied to the separate planning regions in Connecticut. For the moment, however, it will be assumed that, for the particular example under discussion, a set of relative weights has been developed.

The fifth or *Weighting Phase* involves the application of formula I shown earlier. This weighting activity performs two functions: (1) it provides an adjustment mechanism to indicate the true relative importance of the item and (2) it results in the creation of a unit suitable for aggregation. The results of the weighting process are entered into the matrix and serve as the numerical indication of the importance of the particular impact. On the cost or detriment side, this would lead to an artifically weighted numerical expression for the relative impact caused by the loss of recreational land. On the benefit side, the result would show the importance of the additional jobs to the community. While the Weighting Phase is a simple multiplicative operation, it does represent the culmination of all the micro-oriented investigative activity required by the matrix. The unification of both the imputed market figure and community value aspects of each identified impact, yields a single indicator of the estimated overall effect of a specific corridor, as modified by the importance of these aspects to "representatives" of the affected area. If, after the weighting phase the numerical entry in the matrix is large in both an absolute and a relative sense, its importance within the decision-making flow has been justified. Any entry that receives prime consideration in both value systems should exert a stronger force in the evaluation process. Clearly this type of information should serve as the logical base from which a rational decision is made regarding the relative merits of the competing choices.

The *Aggregation Phase* is the sixth step in the evaluation process. This requires the adding up of the individual weighted impacts resulting in a numerical presentation of the total impact. The aggregation process can be accomplished in either of two ways. It is recommended that both be applied, since each yields a different piece of information and serves as a check on the other. The first approach involves the summing of impacts by column with the result serving as an indication of either the burden or benefit attributable to each of the impacted agents. It also yields an indication of the relative division of present vs. future impacts as well as a view of the extent of impact interaction. The usefulness of this approach is apparent in situations where one subcategory of impacted agents feels that it is suffering a disproportionate share of the disruption generated by the project. Following this method, estimates of this disruption can be obtained with appropriate corrective action introduced where necessary.

On the other hand, the weighted subtotals can be obtained by focusing upon impacts and summing across impacted agents. Again, the information is highly useful in that it provides the investigator with an indication of the particular impact factors which dominate the evaluation. With this data, he is made more aware of the areas deserving special attention during construction and which might benefit significantly from marginal changes in design or route choice.

The remaining operation in the Aggregation Phase consists of adding together the weighted tabulations for both the user and nonuser impacts as shown in table 3-5. This addition is also accomplished through the application of a weighted value indicating the relative importance of the dollar impact of each broad classification. As in the previous case the aggregation can be undertaken either row-wise or columnwise with the result of the calculation being the overall net weighted effect of the particular facility being considered. This net result along with the totals representing the overall detrimental and beneficial effects serves as the basis for the operation carried out in the final step.

The final *Evaluation Phase* of the investigation involves the actual comparison of the tabulated results for the various alternatives. This selection process is undertaken in the same way that decision makers have traditionally made final choices within the cost-benefit framework. The evaluation of alternatives with significantly different inputs and results requires the choice of a corridor to be based upon a ratio of project benefits and detriments. The focus of attention in the benefit-detriment ratio is directed toward the "margin" of decision making where the "extra" benefits to be gained from the project with the more favorable net results are compared to the "added" detriment incurred due to the new highway facility. Through this comparison the project with the largest marginal benefit-detriment ratio is selected. This is the corridor to be selected in order to maximize the weighted gain to the area from construction.

It should be emphasized again, in defense of the practical nature of the operation, that the use of the matrix approach does not necessarily mean that the estimated dollar compensation should actually be paid. Nor does it imply that any concrete attempt should be made to establish the budgetary machinery which would be necessary to facilitate the transfer of funds. Both of these acts require professional decisions which are separate from the corridor evaluation and selection process. The matrix simply requires that each impact be considered in its proper relative place within the decision-making flow. The distributional considerations are clearly separate and distinct from the analytical phase of the investigation.

3.8 Discussion of the Row Elements (Impact Factors) of the Matrix

An essential feature of the matrix, as described in the previous section, involves the development of a set of potential impacts that serves as the object of quantification. These impacts identify the ways in which a given corridor alternative can benefit or harm the affected area. As a result, their listing must be complete, with each defined in an operational manner. Obviously, the types of impact will vary between corridor alternatives according to the design and location specifications attached to each. Only by chance would one alternative generate a list of impacts identical with that of another. Furthermore, the existence of a beneficial as well as a detrimental side for each of the impacts further lessens the probability that one fixed list of impacts can be counted upon to appear in every case.

In terms of application, the element of impact variability is itself one of the desirable features of the matrix. As emphasized in chapter 2, the value of an impact approach lies in its ability to abstract the information considered relevant for evaluation from the existing interaction of the subcomponents of the supersystem. By allowing the investigator to begin each evaluation from a flexible point of origin, the matrix implicitly requires him to enter into the impacted area and to work with the people involved to determine which impacts are applicable. Consequently, the quality of the tabulation becomes a function of the extent of the field work and the experience of the investigator.

Upon repeated application, the identification of impacts should become progressively easier as information on past projects is accumulated in what can be considered as an *Impact Data Bank*. If accurately recorded, the data will provide the investigator with the following pieces of information. First, it will yield a list of events that have occurred in the past as a result of highway construction and which can be anticipated to appear on the average in the future. This list would serve as a point of departure for a review of the items, which will determine if they are applicable in the current situation.

Second, if the impacts of past projects are also classified according to the general situation in which they occurred, then the task of the investigator is further simplified. Where conditions are comparable, there will exist a refined list of items available for direct consideration. Again, unique impacts will be encountered with each project under consideration. However, as the refined list expands, the chance is reduced of accidentally missing a key impact through sheer lack of knowledge of the forces at work.

Finally, the availability of a formal account of past investigations will facilitate the development of both descriptions and definitions of the impacts that will enable the investigator to improve his estimates of the value of the impact in the quantification phase. Obviously, if the quantification process is to be carried out so that others may review the results, it is important that the impacts be defined in a way which clearly shows the nature of the item being considered, as well as the lines along which it is being evaluated. The perfection of these aspects of the identification phase should appear as the impacts are repeatedly encountered in varying situations.

The purpose of this section is to offer a sample of thirty general impacts that can serve as the nucleus of the investigative work. These impacts are considered to have both a beneficial as well as a detrimental side and were developed from the body of existing information regarding the potential nonuser effects of highways. The list is not offered as a final accounting of all possible combinations of effects, rather it serves as a first step in the development of the *Impact Data Bank* noted earlier. With this list in hand, the reader can see a practical indication of the types of elements and an organizational format which might be fit into the general matrix form shown in tables 3-3 and 3-4. The individual items further serve as a basis upon which to test several of the assumptions offered in section 4.5 regarding the ability of the investigator to develop a set of relative weights. In effect, what is taking place in this section and those that follow is outlining of the operational validity of the proposed matrix.

3.9 The Source of the Thirty Items

To develop a list of items that possesses an acceptable level of operational realism, four separate avenues of investigation were explored. Each of these was considered to be a valuable source of comment regarding the potential effect of a transport system upon the surrounding environment. Each assumed a slightly different point of view in describing the various aspects of highway impact and, as a result, a broader perspective regarding potential impact areas is developed. There was an attempt to review the source material with three objectives in mind regarding the selection of impacts. First, the impacts which were selected should be generally applicable to all highway construction situations. Although unique situations are probably as numerous *and important*, it was felt that for a test of the procedure the best approach was to select the most obvious effects and organize these into relatively homogeneous groupings.

From this structuring, the second objective could be achieved, namely that the impact being considered has a definition that is recognizable to the average person and which can elicit an intelligent response from those who are asked to perform the weighting. Obviously, without a specific project in mind which yields a given set of outcomes, it is difficult for the respondent to accurately weight the relative importance of the impacts. To compensate for this, it is necessary to supply the respondent with a hypothetical highway example containing a realistic set of potential impacts. Hopefully, he possesses enough knowledge to be capable of relating to these impacts without unnecessary confusion.

Finally, it was felt that the impacts used in the example should have specific relevance to problems which could be encountered within Connecticut. As a result, the examples cited possess a semiurban bias that may not be totally accurate as the investigator operates in states or regions with a rural or agricultural base. However, since the present concern is directed toward the evaluation of Connecticut highway projects, it was felt that the nonuser impacts should be generally applicable.

The first source consulted was the two volume set of reports published by the National Cooperative Highway research program under Project 20-4.[7] These reports were aimed at revealing the attitude of users and nonusers toward both highway construction and automobile use in general. Special attention was also devoted toward the description and prediction of public attitudes toward transport expenditure. Given that the relevant data covers the broad eastern region,[8] it is impossible to divide the information finely enough to view Connecticut or New England separately. However, even with this limitation, the studies are still potentially useful in serving as an initial indicator of construction impacts which may possess general relevance to the corridor evaluation process in Connecticut.

The studies revealed that car owners hold separate attitudes regarding the social value of the automobile, the quality of existing highway facilities and the capabilities of highway planners.[9] In particular, the social role of the automobile was regarded quite favorably by respondents when correlated with any of the common personal characteristics (occupation, age, race, education, etc.).[10] This may be due to a psychological condition known as the "halo effect" whereby people tend to ignore the disadvantages connected with automobiles because of the ego-involvement factor. However, the corresponding attitude towards highway planning and planners was appreciably lower, indicating a general dissatisfaction with the available facilities.[11] Curiously enough, the number of miles driven was found to be independent of both of the above attitudes.[12] In interpreting these results, the conclusion was that the absolute level of use was not an accurate indicator of relative user satisfaction with either the facilities or the planning process itself. Rather, use reflected either the geographic requirements of the population and the lack of adequate substitute modes, or a complete divorce of the concept of vehicle ownership from use of facilities.[13] If the second hypothesis is a more accurate description of the forces governing the response to the questionnaire, this can have significant repercussions upon the nature of public comments regarding the community impact of highways.

This particular attitude of the public, when coupled with the desire to spend more on highways in general, appears to imply that highway planners can expect both a rising tide of vehicle ownership (with its accompanying pressure for added facilities) and an increase in the level of dissatisfaction with the community effects of such undertakings voiced by the very people who initiated the pressure for construction. Obviously, the car owner views the situation differently depending upon the role he is playing at the moment. Education of the public in this matter might result in a more rational acceptance of the cause and effect nature of vehicle ownership and highway building. Many undesirable social impacts may be accepted by informing the people that a choice of priorities must be made. In viewing the response by eastern residents, they appear to be the regional group most dissatisfied with highway planning and planners, while they are also the least enthusiastic about the social role of the auto.[14] Even here, however, those holding a favorable social view of the auto still far outnumber those uneasy about construction.

Those found to be most dissatisfied with the existing transportation system were the younger, higher income, professional, and more educated groups. Given the extent of possible overlap, it can be expected that when united against a particular project, these groups are capable of creating a sizable uproar if planned improvements interfere with their interests or values. In addition, city residents were most concerned with mass transit and were least interested in regarding the auto as an ideal mode. Consequently, undue urban construction of highways can be predicted to create added objections supplemented with public comments directed toward the need to invest in alternative modes.

The level of participation of the above groups in the planning process appeared to vary with their view of the effect of the opinions expressed at hearings. If people felt their comments were taken seriously the participation rate would increase from around 3 to over 50 percent.[15] Obviously, as more attention is paid to the social and economic impacts of highways this will generate both the concern on the part of residents and the motivation for participation by individuals in the planning process. Quite possibly, the resulting exchange of views and information will provide the necessary forum for the education of the driving public with respect to the cause and effect nature of car ownership mentioned earlier. If so, the essential community problems encountered in highway construction might be better anticipated and alleviated.

A second avenue of investigation concerned a general review of the work done by academics in trying to pinpoint the feelings of the general public toward highway construction. Many of the following studies refer to a specific area, or are directed toward the evaluation of a specific aspect of transport impact. As a result, their application to the problems of evaluating impacts in Connecticut should be viewed with care. However, the essential feature of the review is that the following works represent some of the most advanced experiments in the field of value estimation relative to highway planning. Consequently the issue raised by these studies should serve as the logical basis for future work.

The first article investigates the general approach to attitude surveys and the quality of the responses. In discussing problems normally encountered while developing profiles of community views regarding highways (especially route selection), Shaffer feels that most studies aimed at revealing attitudes actually measure opinions instead.[16] She defines *opinions* as unstable views concerning particular situations, while *attitudes* are more basic views, concerned with *abstract* situations. Attitudes are viewed as being more important, since they influence community *values*. Shaffer compares attitudes and opinions as in table 3-6 to highlight to weakness of "opinion studies."

Table 3-6
Opinions versus Attitudes

Opinions	Attitudes
Subject to strong social influence	Social influence is minimal
Easily swayed	Enduring motivating force
Little predictive value	Useful predictive data
Opinions not reflected in action	Indicative of behavior

Source: M.T. Shaffer, "Attitudes, Community Values and Highway Planning," *Highway Research Record* No. 187 (Highway Research Board, 1967), p. 57.

Considering this division, she discusses three possible methods for ascertaining attitudes; the work association technique, the sentence completion technique, and the semantic differential technique, and finds that responses to attitude items are correlated with socioeconomic characteristics. As a practical technique however, these methods seem to involve a tremendous amount of subjective judgment even when large samples are used. Also, there would be considerable difficulty in handling the number of responses necessary from each individual in order to get any indication of attitudes. In summary, it is important when developing techniques for evaluating community impact that the above dichotomy be kept in mind. This should result in the creation of an instrument containing specific items which best measure the topic under investigation.

In order to more fully understand the attitudes toward the abstract characteristics associated with a particular form of transportation, Paine (et al.) tried to develop a research methodology that would accumulate information on consumer feelings regarding movement.[17] They wanted to find what an ideal transportation system would be, from the *user's* viewpoint, and also how closely existing systems come in satisfying this ideal. Using interview questionnaires, the surveys (one in Baltimore and one in Philadelphia) revealed that the major trip purpose distinction is between work and nonwork trips. As shown in table 3-7 most important items were similar for both trip types, with travel time more

Table 3-7
Relative Ranking for Major Trip Factors

	Trip Purpose	
Factors	Work-School	Non-work
Reliability	1	1
Travel Time	2	5
Weather	3	2
Cost	4	4
State of Vehicle	5	6
Unfamiliarity	6	8
Self esteem	7	10
Diversions	8	9
Convenience	–	3
Congestion	–	7

Source: F.T. Paine, A.N. Nash, S.J. Hille, and G.A. Brunner, *Consumer Conceived Attributes of Transportation: An Attitude Study*, p. ix.

Note: The rank of 2 for travel time in work trips actually is similar to the rank of 3 for convenience in non-work trips because of adjustments in the items making up the factors in the survey.

important for the work trip, and convenience and congestion rated higher for nonwork trips.

In addition to the relative ranking shown in table 3-7, most of the factors were rated higher absolutely for the work trip purpose.

The attributes of an ideal transportation system in order of importance were found to be:

1. Reliability of destination achievement (safety, confidence in vehicle)
2. Convenience and comfort
3. Travel time (differing with trip purpose)
4. Cost
5. State of vehicle (cleanliness more important than newness)
6. Self esteem and autonomy (independence rather than pride)
7. Traffic and congestion (in and out of vehicles)
8. Diversions (such as travel companions, radio, scenery)

Those living in the suburbs fairly near the central business district, with one or more cars, high income, a single family home, and who use cars for both trip purposes, tend to rate most of these attributes as important. Those living nearest the central business district in multiple units, with low incomes and education, owning fewer cars, and made up largely of minority groups, tend to rate most of the factors as less important. Persons similar in their characteristics to the first group but living even farther from the central business district, tend to rate most of the factors as not very important at all.

As with the national survey, this study showed an overwhelming preference for the auto over the most likely form of public transportation (see note 7). Travel time, weather protection, self-esteem, and convenience-comfort were the factors on which the auto scored highest. Public transportation was somewhat more competitive for reliability, cost, state of the vehicle, and diversions. However, the auto approaches the ideal more closely than public transportation modes, especially for middle-class suburbanites. This study is useful in considering consumer attitudes towards existing transportation systems. It does not necessarily reveal people's attitudes toward new facilities or the community effects that such projects might have. However, in referring to the characteristics as well as an abstract ideal, it does provide some direction.

The next group of articles turn away from considering the individual solely as a user. Rather, they expand the range of investigation to include the dual role played by the individual as both a user and a community resident. Yevisaker, in his study, sees opposition to highways as a result of the conflict between the desire for mobility and the desire for stability (sacredness of home, etc.).[19] He feels that the use of the majority rule in route location decisions may hurt urban minorities, who get more of the costs and less of the benefits from transportation development. He attributes the source of this problem to the relative imbalance of power. Since more than half of the residents are too young to vote, ghetto areas lack political power. They lack economic power because half of the people are members of households that can never expect to relate transportation values and projects to their own needs in order to gain the benefits from greater mobility.

Continuing this investigation of relative social costs and benefits, Fellman and Rosenblatt discuss the attitudes and social costs of people actually to be displaced by highway development.[20] The particular area studied was found to be occupied primarily by working class people in the lower-middle or upper-lower income class. The social network was tightly knit, with people relying more on walking than on cars.

Since residents had no "natural" institution to work from (except the church) and they often mistrusted and misunderstood the political system, political inexperience and apathy were common.

They found that home owners generally were better informed about the proposed construction, due to their stronger socioeconomic position, and also because they had more motivation to know about the proposed route. Factors often overlooked with respect to displacement are: (1) people invest emotion as well as money into their homes, (2) people lose faith in their ability to keep a home (especially those who have been displaced several times), (3) homeowners have the longest and strongest neighborhood ties, and (4) they often have a strong sense of identity with the community.[21] Renters, on the other hand, tend to watch and wait, since they have fewer economic ties to their dwelling (although they may have strong social ties in the area). Fellmann and Rosenblatt conclude that the lower-middle class simply has a different family structure and relationship with the neighborhood than planners had thought. This points out that the affected area should be studied to ascertain its particular characteristics as a unit.

In a separate study, Dansereau developed findings that tend to supplement the conclusion of Fellman and Rosenblatt regarding low income urban residents.[22] His initial purpose was to compare survey techniques for determining attitudes towards local highway development, planning and zoning practices. Working with suburban and rural subdivisions, he found that attitudes are generally favorable toward planning and zoning, and that they are even more favorable after the distribution of supporting literature informing residents regarding proposed projects. Moreover, he discovered that citizens express general approval of highway improvement, and that the average person agrees with the decision of the community leader regarding route location and design. As a result, approaching community leaders for advice and aid is a valid procedure on the part of highway officials, at least in small nonurban civil subdivisions.

As with the Fellman and Rosenblatt work, Dansereau found socioeconomic characteristics to be an important indicator of attitudes toward planning. In particular, occupation, income, and education are all positively related to the degree of highway used and to personal attitudes toward planning and zoning. Although far from complete, the results are helpful in that they provide an indication of the reactions to highway development which can be expected from nonurban, middle-income persons. This reaction may be caused by the fact that land is more available in semirural areas with fewer people being displaced or even affected in a nonuser capacity by a given project. Obviously, as shown by the national attitude survey and the Fellman-Rosenblatt work, the principal areas of concern are the high income urban-suburban areas and the low-income inner-city core.

In response to the needs of urban minorities, Peattie suggests that planning should shift away from politics and towards expertise in its relationship with low income residents.[23] This can be accomplished by the introduction of "advocacy planning." Public management of urban life can lead to individual feelings of being overwhelmed by the existing political institutions. In recognition of this potential imbalance, the disadvantaged minority should be provided with counselors. These counselors can use their knowledge in specific areas by acting on behalf of the minority to organize it and inform it regarding the implication of specific actions. Most plans contain the necessary elements of group interest which would make the proposal effective. Advocacy planning would be useful, therefore, in political planning to help people make their attitudes felt, especially low-income groups.

There are several problems connected with the use of advocacy planning. Definition of the appropriate client group is one difficult step. The neighborhood may not be a useful unit if it lacks cohesion and agreement among residents. Peattie feels that slums, in particular, are held together only by hostility toward the outside, thus lacking a constructive framework. Natural clients for an advocate planner would be groups which arise in response to a threat. However, these groups may not be representative of the true interests in the neighborhood. Community interests are hard to define since there may be many groups within the community. This may hurt the weakest (who are often the slowest to organize and are typically nonjoiners, afraid of people in positions of authority) when there are conflicting community interests. Advocacy planning is a type of political manipulation whereby localized urban attitudes are organized and presented professionally and can be aired. Its value lies in its attempt to adjust for power imbalances within the community.

These surveys and articles indicate some of the factors which should be considered in evaluation of alternative transportation proposals. They are another step toward trying to determine people's reactions towards highways by ascertaining their attitudes. In effect, they point out which groups can be associated with particular issues and, in general, what highway planners can anticipate as being the dominant areas of concern.

A third source of investigation was a review of the relevant findings derived from a Connecticut-based attitude survey undertaken in 1966.[24] Although the survey does not concentrate specifically on attitudes with respect to transportation, it is an attempt to determine the impressions of individuals regarding the area in which they live. Connecticut residents were asked what they thought about conditions in their towns and the state (features liked, features disliked, problems needing the first improvements, and so on). Information on family characteristics, the history of residential and employment mobility is related to expressed attitudes (i.e., views regarding recreation, leisure time activity, etc.). A total of 4,900 households were sampled over a three-year period by four different agencies. Three sampling techniques were used: home interviews, questionnaires, a hand-out-mail-return questionnaire distributed at a more general home interview, and mail-out-mail-return questionnaire.

The surveys showed that Connecticut people are strongly in favor of low-density housing, especially as family size increases. Accessibility to work is the main determination of the town to be lived in, with nearness to family and friends and the "rural" atmosphere also very important in residential location decision. "Better" neighborhoods are important to families with school age children and to older families. In general, families change their residence (usually to get more space or as a result of a job-change) more frequently than they change jobs (usually to get a better job).

Those happiest with their dwelling live on lots of an acre or more. Next satisfied are those of medium-sized lots (although older families with high incomes often prefer luxury apartments). Slightly over one-third of Connecticut's households moved at least once between 1959 and 1964, with the movers tending to be younger and slightly wealthier than the average state households. Of the group, the most mobile people are married, apartment dwellers under 45, with no children. Half of the moves are to single family homes, with apartments and two-family homes following. These factors indicate that, if income levels continue to rise, the demand for low-density housing may increase. This might result in higher per capita development costs and will also consume a larger amount of available land. A possible alternative might be to provide more amenities on small lots. However, there are small groups who do prefer two-family homes or high-rise apartments. Highway projects that run counter to these findings can be anticipated to generate a high degree of dissatisfaction.

The sampled households were classified by several characteristics, including size of the planning region and income levels. Family composition was factored into six groupings, while housing was divided into ten classifications according to the nature of the dwelling. The results showed that in Connecticut over one-third of all families have a young child of school age, and over two-thirds of the families live in single family houses, with two-family homes and apartments also being popular. As expected, the dwelling type was found to be directly related to income. Towns also divided into five categories based upon population density and growth rate. It was found that more than half of the families live in low and middle density, rapid growth towns.

Town features most liked by residents (in decreasing order of preference) are:

1. The suburban-rural atmosphere
2. The town's general appearance
3. Recreation opportunities
4. Convenient location
5. Public education facilities
6. Shopping facilities

This varied somewhat, with the suburban-rural atmosphere important in smaller, while large towns emphasized parks and recreation. Convenient location was important in medium-density towns, and with rapid growth areas indicating their satisfaction with their shopping facilities. The data showed a consistent relationship between the number of people who like their town and both the density and the growth rate of the town.

Town features most disliked are:

1. Inadequate recreation and entertainment opportunities
2. Poor streets and highways
3. High taxes
4. Poor rail and bus service
5. Poor government services and facilities

Rapid growth towns particularly complain about inadequate recreation and high taxes. Slow growth towns are concerned with slum areas and lack of urban renewal. Low-density rapid growth areas seem to dislike poor shopping facilities and inadequate planning and zoning. The major town problems are most often stated to be redevelopment with high taxes, poor education, lack of planning, need for attraction of industry, and lack of provision of adequate streets and parking also being mentioned. Schools, haphazard expansion, and lack of planning are considered to be the major problems in rapid growth towns. Providing job opportunities is a problem in small towns. Clearly, a town's problems change as the town grows and ages.

In general, town recreation facilities are appreciated when they are good, and missed when they are not available. Public recreation facilities seem to be somewhat better than local entertainment facilities. Middle-income people are more satisfied with recreational opportunities than other income groups. Old, large towns are usually proud of their recreation.

Schools, streets, and highways are often rated as satisfactory. Streets and highways are more of a problem in older, slow-growth towns of medium and low density. However, the surveys gave no indication whether the problem is poor road quality, traffic, or lack of high speed facilities. Only 9 percent of Connecticut residents use public transportation, and most of those are critical of it. Slow growing medium and low-density towns rate bus and rail service as poor. The studies imply that Connecticut residents are rather concerned about provision of public transportation. This probably would lead to higher residential densities than they prefer, but the availability of transit perhaps would be worth the added disadvantages, since it would in turn help preserve open spaces and maintain the rural-suburban atmosphere.

The state features most liked are: (1) general appearance, (2) highways, (3) recreational facilities, (4) the rural-suburban atmosphere, (5) climate, geography and size, and (6) job opportunities. On the other hand, state features most disliked are: (1) highways, (2) taxes, (3) government policies, and (4) political atmosphere. In general, the most important state problems are considered to be attracting industry, improving highways, taxes, and general growth problems. People do not seem to differentiate between town and state responsibility for problems, with the exceptions of highways and recreation. It is somewhat surprising that highways, rated second on the "liked" side, is also the feature rated highest on the "disliked" side. People seem ambivalent toward certain aspects of the highway system. It may be that many use Connecticut roads so often that they are aware of needed changes.

State residents are also concerned about Connecticut's economic base, realizing that additional job opportunities may be provided through attracting industry, and that this will broaden the tax base. There is often a conflict between desired service levels and the taxes needed for them. This is especially so in low-density, rapidly-growing towns where services have not been provided and in slow-growing, high-density towns where modernization and redevelopment of older areas is needed.

Another facet of the analysis concerns attitudes toward leisure time. In order, Connecticut residents indicate that they prefer to spend their leisure time: (1) reading, (2) watching television, (3) gardening, (4) hunting and fishing, (5) outdoor sports or (6) on hobbies. On weekdays, most leisure time is devoted to around-the-house activities, while on weekends more time is spent on preferred activities, more often outside the home. Comparison of actual with preferred leisure time activities shows a fair correspondence, except that people spend more time watching television than reading. Higher income people expect more recreational services, and per capita recreation usage rises with increasing income levels. Connecticut residents want both town and state swimming and park facilities, along with better cultural and entertainment opportunities.

In comparing attitudes among regions, as shown in table 3-8, the following variations are more interesting than the regularities. Most regions' residents say nearness to job and friends is the main determinant in town selection. Rural areas, however, feel that the rural character is a major factor. In more urbanized regions (southern Connecticut) the type of residential development is more important. Every region, except greater Bridgeport, chose the suburban atmosphere as the feature most liked. Southern Connecticut and the New Britain-

Table 3-8
Regional Attitude Variations

Planning Region	Attitudes: Reason for Selecting Towns	Town Feature Most Liked	Town Feature Most Disliked
Northwestern Connecticut	Near Job	Suburban-Rural Atmosphere	Poor Shopping
Torrington Winchester	Near Friends Near Job	Suburban-Rural Atmosphere	Recreation and Entertainment
Capitol	Near Job Near Friends	Suburban-Rural Atmosphere General Appearance	Recreation & Entertainment; Rail & Business; High Taxes
Windham	Near Job Near Friends Rural Character	Suburban-Rural Atmosphere	Recreation and Entertainment
Northeastern Connecticut	Near Job Near Friends	Suburban-Rural Atmosphere	Poor Shopping Rail & Bus
Southeastern Connecticut	Near Job Near Friends Rural Character	Suburban-Rural Atmosphere	High Taxes; Streets & Highways; Government
Connecticut River Estuary	Near Job	Suburban-Rural Atmosphere	Poor Shopping; Poor Management of Planning & Zoning
Midstate	Near Job Near Friends Rural Character	Suburban-Rural Atmosphere	Recreation and Entertainment
New Britain-Bristol	Near Job Near Friends	Park & Recreation Suburban-Rural Atmosphere	Streets & Highways & Bus; Appearance
Central Naugatuck Valley	Near Job Near Friends	Suburban-Rural Atmosphere General Appearance	Streets & Highways Recreation and Entertainment
Housatonic Valley	Near Job Near Friends Rural Character	Suburban-Rural Atmosphere	Rail & Bus
South Central Connecticut	Near Job Near Friends Type of Residential Development	Suburban-Rural Atmosphere Parks & Recreation	Streets & Highways Recreation and Entertainment
Valley	Near Job	Suburban-Rural Atmosphere	Rail & Bus
Greater Bridgeport	Near Job Near Friends Type of Residential Development	Parks & Recreation	Poor Government Lack of Urban Renewal High Taxes
Southwestern	Near Job Near Friends Type of Residential Development	Suburban-Rural Atmosphere Parks & Recreation	Traffic Recreation and Entertainment

Table 3-8 (cont.)

Planning Region	Attitudes			
	Most Important Town Problem	State Feature Most Liked	State Feature Most Disliked	Most Important State Problem
Northwestern Connecticut	High Taxes	Highways	Highways High Taxes	Attract Industry Highways; Taxes
Torrington Winchester	Providing Job Opportunities Lack of Planning Inadequate Schools	General Appearance	Highways Climate	Attract Industry Highways
Capitol	High Taxes	General Appearance	Climate Highways	Attract Industry Highways; Taxes
Windham	Inadequate Schools Attract Industry	Highways	Climate Highways	Attract Industry Cost of Public Education
Northeastern Connecticut	Lack of Planning Inadequate Schools	Highways	Highways Job Opportunities	Attract Industry
Southeastern Connecticut	Gov.'t Services Attract Industry	Highways	Highways	Attract Industry
Connecticut River Estuary	Lack of Planning	General Appearance Highways Recreation	Highways High Taxes	Growth Problems Highways Attract Industry
Midstate	Gov.'t Services Inadequate Schools	Highways Recreation	Highways	Industry Highways
New Britain-Bristol	Redevelopment	Recreation	Highways	Highways
Central Naugatuck Valley	Inadequate Response	General Appearance	Recreation	Attract Industry
Housatonic Valley	Lack of Planning Inadequate Schools	General Appearance	Highways	Highways
South Central Connecticut	Redevelopment High Taxes	Highways General Appearance	Highways Gov.'t Policy High Taxes Politics	Attract Industry Highways
Valley	Redevelopment Inadequate Schools	Highways	Highways High Taxes	Attract Industry
Greater Bridgeport	Redevelopment	Recreation Highways General Appearance	Highways Gov.'t Policy Politics	Attract Industry Highways
Southwestern	Redevelopment	General Appearance	High Taxes Gov.'t Policy Politics	Attract Industry Growth Problems

Source: Alan M. Voorhees & Associates, Inc. *Analysis of Survey of Personal Attitudes,* 1966, figures 8-14, pp. 115-130.

Bristol region also rate their parks and recreation highly. Poor government services are a problem in the greater Bridgeport and south-central areas, while the Connecticut River estuary region feels the need for greater planning. Redevelopment is a problem for older towns, while schools and government services are often considered inadequate in the less-urbanized regions. Highways are important to all regions, with the Housatonic Valley and New Britain-Bristol regions wanting to attract industry. Growth problems are major in the southwestern and Connecticut River estuary regions.

The survey indicates that a significant degree of diversity in attitudes exists among regions. This apparent variation should support the following hypotheses regarding the row elements of the matrix: (1) different items will be included in the evaluative matrix as the investigation moves across geographic subdivisions; (2) the same item will have a varying importance depending upon the region being investigated; (3) the importance will also vary depending upon whether the same impact is beneficial or detrimental. Obviously, the attitudes revealed in this survey should serve as a further indication of the highway impact factor considered important by the people of Connecticut.

The fourth and final source of investigation was actually a combination of two review areas. The first involved a reading of the transcripts of the hearings on existing and proposed federally-aided highway projects in Connecticut.[d] From these hearings, a series of items was developed which indicated the major concerns of residents in the impacted areas. The second review source was the set of directives issued by the Bureau of Public Roads regarding the factors which must be considered by state highway departments when building roads which are financed in part with federal money. It was felt that by approaching the topic from the direction of expressed and anticipated concern, the final link could be provided for a practical list of items to be used in our own survey.

The hearings review pointed out the following instances of concern. First, people were very anxious regarding the potential physical impact of the facility on the community. In particular, accessibility was of primary interest in relation to both vehicle and pedestrian movement. The fears expressed regarding movement were generally of two types: (1) that entrance and exit facilities would place added traffic on already over-crowded city streets, and that (2) the highway would act as a barrier to the flow of traffic from one side to the other. This anxiety was compounded by the notion that the highway would add to the existing level of pollution (visual, noise, water or air) already in the area. These dual physical aspects of accessibility and pollution were concerns in both the short run and the long run as spokesmen felt that even the temporary difficulties caused during the construction period deserved significant attention. This concern over physical impact centered mainly on the cost side with very few speakers rising to comment on the improved traffic flow and reduced congestion which could be anticipated from the project. Although such helpful effects usually occur, the beneficiaries are slow to step forward and acknowledge them.

[d]The hearings reviewed included those in Interstates 89, 91, 95, and 291.

Economic feasibility was a second source of concern with the majority of comments focusing on the impact of the facility on land values in general with occasional reference to specific areas. Homeowners were especially motivated to point out that either acquisition costs were inadequate, or that property which would fall in value because of the highway should also be purchased. They questioned whether these added costs were properly considered along with user and construction charges. Moreover, speakers described potential changes in patterns of land use that could generate a sizable indirect cost burden to be borne by community residents. Again, few people spoke of the potential benefits to be gained economically for an area. Except in rare cases, where the road was planned or justified on the basis of opening the area to economic development, those experiencing the benefits failed to describe them.

Other commentators asked for a more in-depth review of the personal aspects of disruption. The issue of relocation was especially important in urban areas where the difficulties of obtaining replacement housing were prominent. Also, the question of equity arose in some cases where the benefiting and suffering groups appeared to have little overlap. Here, people were concerned with the measurement of impacts and their adequate compensation.

In summarizing the comments made at hearings, it appears that most were motivated by some aspect of personal concern or involvement. In only a few instances were issues raised which could be interpreted as being community-wide in nature. However, in these cases, the level of vocal opposition reached surprising proportions. It would be difficult to judge the true cause of such reactions; however, the number of people affected, the emotional nature of the impact, or its severity could be considered controlling factors. In retrospect, it is somewhat surprising that more people do not speak out in favor of highways. Several hypotheses can be offered in explanation of this phenomenon: (1) the benefits are not real; (2) the benefits are offset by greater detriments to the same people; (3) the beneficiaries are apathetic; (4) the beneficiaries feel that the highway department is an effective lobby and needs no support from individuals. Which, if any, of these actually explains the observed phenomenon is a question open to future consideration.

The use of Bureau of Public Roads directives as an informational input in the development of the matrix has a special significance. Within recent years, the United States Congress and the bureau have become increasingly aware of the connection between community development and highway construction. In recognition of this link, Congress initially passed the 1962 Federal-Aid Highway Act, which called for a closer coordination between highway construction projects assisted in part by federal funding and local planning activities. The goal of the law was to ensure that the two activities complimented and were consistent with each other. In the enforcement of this provision throughout the country, the record has been spotty at best.[25] Interpretation of the relevant passages has favored the actions of state highway departments so that construction delays are eliminated and planning control remained in the hands of state highway officials.

To partially correct for this problem, Congress passed a second community oriented law, the Federal-Aid Highway Act of 1968. The act, among other items, legally bound the state highway departments to consider the social, economic, and environmental aspects of highway impact during the hearings process. The results of these hearings must in turn be submitted to the Bureau of Public Roads for review and approval.

To assist in this review, the bureau issued a list of twenty-three (nonexhaustive) items which should be used as a starting point in the investigation process.[26] This list of impact items, as shown in table 3-9, has served as a basic indication of preliminary federal concern on the matter, and has been developed

Table 3-9
Twenty-Three Potential Social, Economic, and Environmental Impacts of Highway Construction

1. Fast, safe and efficient transportation
2. National defense
3. Economic activity
4. Employment
5. Recreation and parks
6. Fire protection
7. Aesthetics
8. Public utilities
9. Public health and safety
10. Residential and neighborhood character and location
11. Religious instructions and practices
12. Conduct and financing of Government (including effect on local tax base and social service costs)
13. Conservation (including erosion, sedimentation, wildlife and general ecology of the area)
14. Natural and historic landmarks
15. Noise, air, and water pollution
16. Property values
17. Multiple use of space
18. Replacement housing
19. Education (including disruption of school district operation)
20. Displacement of families and businesses
21. Engineering, right-of-way, and construction costs of the project and related facilities
22. Maintenance and operating costs of the project and related facilities
23. Operation and use of existing highway facilities and other transportation facilities during construction and after completion

Source: "Policy and Procedure Memorandum 20-8," U.S. Department of Transportation, Federal Highway Administration, Bureau of Public Roads (January 14, 1969), p. 2.

with modifications as one component of the matrix row elements. Unfortunately, as presented by the bureau, the items lacked a cohesive organizational structure as well as operational definitions. Furthermore, the accompanying directive lacked instructions as to how the investigation of these items was to be carried out. As a result, a highly selective process was used to sort out the items considered relevant for Connecticut and to develop meaningful definitions. Given this manipulation it still should appear evident, after reviewing the following thirty hypothetical impacts, that the general spirit and intent of the original directive has been maintained.

3-10 The Tentative Impact Items

From the preceding four sources the following list of thirty items was developed. The list is only suggestive of the types of factors which might be included in the Impact Data Bank. All the items can be considered from either the beneficial or the detrimental side and definitions as well as possible methods for dollar quantification are provided in outline form. A more formal list of items with their appropriate operational notation can only be developed as the matrix is applied to actual situations. Finally, the items are arranged in separately defined categories according to recognizable characteristics. This organization, while tentative, does provide an indication of one possible way in which the items can be grouped for practical analysis and review.

I. Impact Category I: *Aesthetic Impact Factors.*

This group contains nonuser or community impacts of highway construction which have an effect on the senses.

A. *Visual quality of the highway itself.*
  1. Definition–The extent to which the architectural design of the highway itself is visually satisfying.
  2. Dollar Measure–The dollar amount actually spent or which should be spent to make the structure visually acceptable.

B. *The visual compatibility of the highway with the existing background.*
  1. Definition–The extent to which the highway blends with the natural or man-made background.
  2. Dollar Measure–The amount of money actually spent or which should be spent to achieve compatibility.

C. *The aesthetic value of the land used for the right-of-way.*
  1. Definition–Sensual worth of the right-of-way land.
  2. Dollar Measure–The amount of money that must be spent to generate a comparable level of sensual value.

D. *The short-run aesthetic impact on the community caused by temporary construction facilities.*

1. Definition–Temporary sensual changes in the level of environmental quality (air, water, noise or soil) brought about by the construction process.
2. Dollar Measurement–Amount of money needed to return the environment to the preconstruction status-quo.

E. *Permanent aesthetic impact on the quality of the environment.*

1. Definition–Long-run sensual changes in the level of environmental quality (i.e., levels of air, water, noise or soil pollution).
2. Dollar Measurement–Amount of money needed to return the environment to the preconstruction status-quo.

II. Impact Category II: *Economic Impact Factors.*

This grouping contains nonuser community impacts of highway construction which alter the monetary standard of living of area residents.

A. *Change in the number of available jobs.*

1. Definition–Changes in the number of jobs or their distributional classification.
2. Dollar Measurement–Changes in the purchasing power or level of income caused by the changes in employment or variations in pay on existing jobs.

B. *Change in the number of welfare recipients.*

1. Definition–Fluctuations in the recorded number of persons receiving welfare payments.
2. Dollar Measurement–Changes in the total dollar amount paid to recipients.

C. *Changes in the value of area property.*

1. Definition–Change in the assessed value of land and and buildings.
2. Dollar Measurement–Dollar amount by which the value of land and buildings has altered.

D. *Short run economic effects caused by the planning or construction phases of the highway.*

1. Definition–Temporary changes in items A, B, C and E or monetary effects caused by uncertainty or locational changes.
2. Dollar Measurements–The short run dollar value of the identified changes.

E. *Change in the level of income.*

1. Definition–Change in the amount or purchasing power of the income.
2. Dollar Measure–The monetary fluctuation in income levels.

III. Impact Category III: *Political Impact Factors.*

This group considers nonuser community impacts of highway construction which alter the way the community is affected by or participates in the actions of government.

A. *Changes in Municipal Services.*

1. Definition–An alteration in either the quality or quantity of municipal services provided per capita.

   2. Dollar Measure—The *dollar* amount needed to account for the change in service.
 B. *Change in the level of participation in government decision making.*
   1. Definition—Number of times people participate in government related activities.
   2. Dollar Measure—The amount of money that would have to be spent to generate the appropriate change.
 C. *Change in the financial capabilities of local government.*
   1. Definition—Change in the size of the tax base of the community.
   2. Dollar Measure—Alteration in the tax revenue attributable to the tax base change.
 D. *Change in the level of community security.*
   1. Definition—Change in the number or type of crimes committed in the community.
   2. Dollar Measure—The amount of money that would have to be spent to compensate for the change.
 E. *Change in the level of community satisfaction with government actions.*
   1. Definition—Change in the percentage of area residents who are hostile or favorably disposed toward government actions.
   2. Dollar Measure—The amount of money that would have to be spent to either correct or account for the change in attitude.

IV. Impact Category IV: *Land Use Impact Factors.*

This group considers nonuser community impacts of highway construction which alter either the existing or desired form of land use.

 A. *Changes in the desired pattern of land development.*
   1. Definition—Change in the percentage of land in the community available for potential use in each of the principal land use classifications (zoning).
   2. Dollar Measure—The amount of money needed to correct or account for the observed changes.
 B. *Change in the number of business firms.*
   1. Definition—Change in the percentage number of firms in each of the types of business organization (i.e., commercial, manufacturing, distribution, etc.).
   2. Dollar Measure—The amount of money that would have to compensate for the change in the number of firms.
 C. *Change in the amount of open space recreational, or conservation level.*
   1. Definition—Change in the amount of land actually devoted to land preservation, personal activity, or ecology.
   2. Dollar Measure—The amount of money that would have to be devoted to these purposes to account for the change.
 D. *Change in the availability of living units.*
   1. Definition—Alteration in the number of single homes and apartments in the area in specific income classes.

2. Dollar Measure–The amount of money needed to bring about the recognized change in living unit availability.

E. *Change in the number of historic sites.*

1. Definition–Change in the number of sites considered as having a link with past events which could interest viewers.
2. Dollar Measure–The amount of money spent on viewing the site.

V. Category V: *Health and Safety Impact Factors.*

This group considers nonuser community impacts of highway construction which affect the level of physical well-being of part or all of the community.

A. *Change in the accessibility of emergency facilities.*

1. Definition–A fluctuation in the average length of time needed to receive these services in response to a call for assistance.
2. Dollar Measure–The amount of money needed to account for this change in response time.

B. *Change in the compatibility of the area with national defense needs.*

1. Definition–Change in the level of national or area security caused by an alteration in the flow of necessary items during periods of crisis.
2. Dollar Measure–The amount of money that must be spent to account for the flow.

C. *Change in the level of environmental pollution that is a threat to general health.*

1. Definition–Increase in the volume of pollution above federal or state safety levels.
2. Dollar Measure–The amount of money that must be spent to account for the rise in pollution.

D. *Change in the level of community stress.*

1. Definition–Change in the level or type of tension felt in the community.
2. Dollar Measure–The amount of money needed to account for a comparable change in tension levels.

E. *Change in the level of safety on Routes that are adjacent to the New Highway.*

1. Definition–Fluctuation on the level or severity of accidents on parallel or feeder roads.
2. Dollar Measure–The amount of money needed to correct or generate the observed change.

VI. Impact Category VI: *Social-Psychological Impact Factors.*

This group considers nonuser community impacts of highway construction which alter the existing patterns of personal or group relationships.

A. *Change in the level of community cohesiveness.*

1. Definition–Change in the community's ability to agree and act upon generally accepted set of goals.
2. Dollar Measure–The money needed to generate the observed change in cohesiveness.

B. *Change in the number of important daily contacts of a personal or business nature made within the community.*

1. Definition—Change in the number of times each agent in the community makes a contact with another preconstruction agent.
2. Dollar Measure—The amount of money that need be spent to account for the observed change in contracts.

C. *Change in the level of neighborhood stability.*

1. Definition—The type and volume of gross resident-flows in and out of the community.
2. Dollar Measure—The amount of money needed to account for the observed flows.

D. *Change in the interaction of the community with surrounding communities.*

1. Definition—Change in the physical accessibility of other communities which cause a fall in intercommunity contacts.
2. Dollar Measure—The amount of money needed to account for the change in physical accessibility.

E. *Change in the number of community related contacts.*

1. Definition—Fluctuation in the number and type of important religious, recreational, educational or community group contacts.
2. Dollar Measure—The amount of money needed to be spent to generate the observed change in contacts.

Part II: Application of the Methodology: A Case Study of Connecticut

4 The Development of a Set of Weights

4.1 Alternative Approaches for the Selection of Those Persons Capable of Performing the Weighting Function

Several attempts have already been made to develop the necessary trade-offs between impacts by assigning a system of weights to the objectives used in the analysis.[1] These attempts introduce an added dimension to highway evaluation, since the derivation of the weights themselves becomes a key element in the decision-making flow. As noted in section 3.7 the weights result in each impact factor being adjusted by the importance of the item within the hierarchy of community objectives. For instance, due to the weighting factor, an imputed compensation of $1000 for the added cost of increased air pollution might not require a comparable $1000 increase in the estimated market benefits derived from a rise in the overall community skill level. By assigning weights of say 5 and 25 respectively, the members of the community would be indicating that the absolute level of cost due to air pollution would have to increase by a dollar factor of five before its decision-making importance was equal to a onefold rise in skill level. As long as this trade-off ratio prevails, those alternatives will be chosen whose weighted contribution to educational level exceeds the weighted value of the loss that the community attributes to air pollution. On the other hand, alternatives which apparently possess a significant level of market benefit as the result of an anticipated reduction in the level of air pollution, might be overwhelmed by a much smaller dollar compensated loss in skills if the above weighting ratio holds on the opposite side of the ledger. Obviously great care must be exercised in developing a set of community values by which the estimated market impact is to be weighted.

Ideally, the matrix weights should be a mirror image of the community goals, so that any controversy concerning the selection of an alternative corridor can be settled to the satisfaction of all by appeal to the decision-making matrix. The first set of difficulties arises when the origin of the community weights is considered. Who should be responsible for developing the broad outline for community action? What individual or group of individuals is most capable of evaluating the future of the community? Which elements are considered to be of prime importance to the community? No single answer to these questions exists due to limitations and imperfections in the decision-making process. It should be noted that even in an ideal situation the answers derived from the different groups may not be the same. What is needed is an explanation of the possible choices and a selection of one which best fits the needs and restrictions of a particular situation.

The most obvious method of determining community goals, as mirrored through the accompanying weighting scheme, is by asking the residents to state their preferences through some form of voting procedure. The voters acting as a body might be asked to rank or select a certain number of goals from a prepared list which would hopefully represent the universe of possibilities. Alternately, they could be instructed to write down what they feel are the most vital community goals in a way which describes their nature and indicates the order and magnitude of their importance. Although the approaches follow the democratic tradition, they both possess some serious drawbacks. First, on a general level the dollar cost of such an undertaking would be significant. To organize, administer, tabulate, and evaluate the election results might well involve a cost that most communities would not be willing to bear. The transfer of scarce dollar resources from the provision of community services to the funding of a voting process may violate the very goals revealed by the election. Furthermore, considering the willingness of the public to participate, and certain voting peculiarities which limit the ability of particular groups to vote it can be expected that the results will not reflect the true goals of 100 percent of the community.

Second, if a prepared list is presented to the voters, it is quite possible that all the relevant goals would not be included or the description of each goal may not convey the same meaning to all participants. In either case, the quality of the list and its usefulness as an evaluator is diminished. Finally, it has been demonstrated that a possible inconsistency can result in the voting process which nullifies the conclusions.[2] Known as the "voting paradox" it points out that ranking preferences (i.e. community goals) by voting can lead to intransitive results since the outcome depends partially upon the order and way in which the alternatives are presented. Clearly a prepared list of goals is not desirable. If, on the other hand, the alternate voting plan of having voters write down in their own words the goals they feel are most desirable is adopted, then problems of a different nature appear. At the precise moment of voting, some individuals may not be able to list all the goals considered to be important. By omitting certain elements or failing to rank them in their proper order, the voters unconsciously alter the future plans of the community. Furthermore, interpretation of hand-written goal descriptions can lead to confusion over terms, and uncertainty. Again, the ideal of having the results mirror the attitudes of the residents is placed in doubt.

There is one final difficulty that must also be faced if voting is to be the means by which community goals are ascertained. The problem revolves around the point that there are two classes of goals which are not adequately represented when mature voters act individually. The first class involves public goods which require the community to act as a body in order to adequately recognize and provide for them. Items such as defense and public health measures would fall into this category. It is doubtful that a large number of voters acting independently would place these items very high on their list *if* they were included at all. However, it is clear that without these two factors being included as important community goals, the possibility arises that the continued existence of the community itself might be threatened. The second class of goals usually omitted or understated by individuals concerns the sufficient consideration of elements which bear fruit at some distant future date. The desire for the present gratification of individual consumer needs which underlies much of our activity would have to be partially suspended by voters if scarce resources available today are to be transferred toward the production of consumer goods at a distant point in time. Normally, individuals acting independently are reluctant to make such a sacrifice on a scale which will adequately provide for the future. For example, forests not cut down today have a high opportunity cost in forgone consumer products. However, if all trees are turned into wood products today, there will be no shaded park lands or conservation areas for tomorrow's citizens. In medical research and education too, the voluntary provision of today's resources for the benefit of future generations is necessary. Obviously, a community conscience is needed here to guide the allocation of resources. People acting individually have a tendency to think only of their own needs and consequently a rational consideration of future oriented goals requires the community to act as a body.

If direct voting by the totality of community residents is rejected as a viable alternative due to the objections raised above, it may be possible to partially correct these faults by engaging in a weighting based on the techniques of random sampling. The limited nature of the sample is no more harmful than the limited voting turnout, while the reduced cost and the presence of a trained impartial interviewer enhances the practical side of the results. In dealing directly with the respondent at length, it becomes possible to probe in greater detail the consistency and importance of the goals stated. Furthermore, other valuable information can be acquired which allows a cross-classification of weights by personal characteristics and group tendencies. This information can be quite valuable in determining the attitudes of neighborhood groups within the larger community. Some of the faults mentioned above still remain, however, especially in the area of providing resources for the future, and a new diffficulty is introduced in the form of sample bias or sampling error. At best the survey technique can only serve as a starting point for the development of a comprehensive system of weighted community impacts.

If the community acting as a whole is unable to derive goals and establish weights in an unqualified manner, then it might be that if the task is to be performed at all, it will have to be undertaken by some group or individual acting for the community. Community goal setting by a single leader is rejected for the most part since it fosters a neglect of community participation, and as shown by past events, it can lead to disastrous results caused by the capricious whims of a single man. Consequently a group should be selected which, in its composition, reflects and recognizes those goals that are best for the community in both the short and long runs. Three groups have commonly been used in different studies as the source of goals and weights; political leaders, community representatives, and professional planners. Each group separately possesses decision-making characteristics which makes it well-suited to the task.

The political leadership is the most obvious first choice since they have already been selected by the community as the organizers or administrators who are best able to handle the governing of current affairs. If present tasks are being adequately performed, then there is some justification in assuming that the weights can also be satisfactorily set by this group. The recurrent need to obtain approval for past actions and future plans at election time allows the voters to correct decisions which they feel are in error. The major feature, therefore, of this group becomes responsible consensus. Political leaders are constantly being fed current information on the attitudes of those they serve, and this fact, when combined with professional insight and experience, can lead to useful judgments concerning the weights.

The main practical difficulty in allowing this group to exercise the decision-making power resides in the strength of the link to actual community goals through the voting process. Admittedly, the political group can abuse the power extended to it by developing weights which favor its own interests as opposed to that of the community. Correction of this problem through the voting process is slow, at best. However, cases of flagrant abuse are rare and for the most part elected officials work honestly for the betterment of the community. For these groups the link as presently conceived appears to be strong enough. In fact, the degree of flexibility that is permitted the honest politician is necessary if he is to perform his job fully. What is usually overlooked in discussing the decision-making role of the political leader is that often he must make unpopular decisions.[3] As shown earlier, provision must be made for certain goals which would be either rejected or given lesser weights by individuals. In allowing for these goals, political leaders are properly performing their job and hence must be given some slack in the reins if the good of the community as a whole is to be considered.

In place of goal setting by politicians it is sometimes felt that community leaders should exercise this responsibility. As the spokesmen for group interests in the neighborhood, they are usually cited as having the ability to interpret and articulate the community viewpoint. Again, such groups should cover all of the area's residents. By allowing group leaders to set the weights an advantage is gained in that they are less likely to present a myopic view of the future which can occur with persons acting individually. In the weighting process, each group would present its own collective opinion which would tend to offset the goals of others necessitating some form of group compromise and consideration of society as a whole.

Part of the limitations of this plan center on the fact that these community leaders are not elected representatives who possess some realistic basis for their claim of leadership. They are often self-appointed or nominated by other individuals within the group. The possibility is always present that these leaders do not represent the totality of community interests or that they lack an accountable responsibility to the groups they are to serve. If these criticisms are in accurate reflection of the leadership situation, it becomes exceedingly difficult to determine which leaders of what group are to be included in the planning process. Surely, due to practical limitations, all those who profess to possess leadership credentials cannot be included. However, in denying certain persons or groups representation, the risk arises that those excluded really did have something useful to contribute and that their interests will have to be considered at some later date.

A second difficulty resides with the selection of a source to establish the organization and pick the community leaders who will serve as members. Often an elected official (i.e. mayor, governor, legislative committee chairman, etc.) is empowered to make such a selection. Because of this appointment process, it is quite possible that the resulting attitude or composition of the community leaders does not truly reflect the feelings of the community at large. Obviously the choice of who should do the selecting is as important as who should be selected.

Finally, another possible problem area involves the fact that in most cases community leaders lack both the time to devote to the weighting task and the professional ability to express goals or weights in a usable manner. The discussion, evaluation, and formalization of a comprehensive set of community goals requires a depth of involvement that a majority of the community leaders are unable to provide on a part-time basis. As a result the task of establishing goals and weights falls to a select few who can engage in this lengthy activity, with the main body serving only as a source of approval. To the extent that such a division of activity occurs the chances for inaccurate results also increases. Furthermore, even if a verbal expression of goals and weights is possible, it may well be that the statements are operationally useless. Goals such as "improving the quality of life" while legitimate statements do require a more extensive explanation before they can be used to evaluate specific projects. A lack of professional training can lead to unnecessary confusion in this area.

Partly as a solution for the above limitation of civic leadership, it is often suggested that professional planners be designated by the proper authority as the weighting agent for the community. As professionals in the field they possess a working interest in the community. Consequently, they often appear as impartial interpreters and arbitrators of the community goals. Their skill and insight usually generate a detached perspective that allows for operationally useful descriptions of goals and systems of weights. The use of planning groups composed of experts in different fields creates a multidisciplined approach which is most likely to cover all relevant objectives. At the same time, the professional group approach leads to consideration of goal interaction which is often ignored by political or civic groups. A limitation of the use of professional groups is that they are not in any meaningful way immediately accountable to the public for their decisions. The detachment that develops decision-making perspective also creates a certain amount of coldness (often labeled "objectivity") to individual or group pleas. Such actions if carried to an extreme can lead to serious distortions especially in the weighting system. Planners are more likely to use their own judgment in allocating resources for use by future generations. Strict zoning regulation, conservation, social interaction, and other factors often receive a level of attention which exceeds the amount considered justified by the community acting as a body. Decisions by professional planners, especially if they are assigned control over project evaluation, can lead to serious conflict with established community interests.

From the analysis of group goal setting and weighting presented above, it is evident that each approach possesses desirable as well as undesirable elements. The ideal method would attempt to accentuate the beneficial goal-setting characteristics of each group while minimizing the impact of the limitations. This is usually accomplished by allowing all three groups to have some role to play in the decision-making process. The common decision-making format consists of a composite group made up of political and civic leaders who discuss the area goals with advice being sought and considered from academic authorities or professional planners. In addition, depending upon the size of the goal-setting project and the availability of funds, an outside consultant is often employed as the full-time, experienced, analytical staff of the committee. In performing the task in this way (often with the aid of public hearings to acquire individual opinions) the results are usually professionally constructed, operationally impartial, responsive to community needs and capable of being directed in an overall manner by the community residents. In following this approach an honest effort is put forth to obtain the most accurate view of the community attitude.

Before advancing to a discussion of the actual establishment of weights and goals for the matrix outlines in the remainder of the chapter, it is important to note that this goal-setting process must be a continuing effort. To be truly effective, periodic review and updating of the community's views, since these views fluctuate over time (if only in a marginal sense). The existing population alters its attitudes in response to changes in both its socioeconomic characteristics and its level of maturity or sophistication. These bring about a new perception of the role to be played by man and the community. Moreover, as mobility increases this usually adds a reinforcing effect to goal change as new persons with different backgrounds and values move into an area. It is quite likely that even if the list of goals remains the same, the weights attached to them will alter over time due to the above reasons. Obviously, a continuing watch must be kept in this area if alternative projects are to be evaluated accurately.

4.2 The Nature of the Group Selected

This section contains an analysis of the composition and reasons for selecting the group used to establish the relative weights, for the thirty hypothetical impacts, developed in section 3.10. It was felt that the only proper way to perform this function was by approaching representatives of the public to solicit their cooperation in the weighting activity. This public participation was encouraged on a controlled and highly selective basis, which would allow for a testing of the hypotheses that were offered in support of the operational nature of the evaluative framework in section 4.4.

It was felt that the most efficient method for obtaining the desired results was on a regional basis, working with the cooperation of the individual regional planning agency. This regional organization was highly convenient for the weighting process, since it possessed existing contacts with individual representatives from towns throughout the state. Thirteen out of the fifteen planning agencies were in full operation and agreed to cooperate with the study. The advantages of the regional approach are threefold. First, it allowed the research group to cover approximately 93 percent of the state's population, and approximately 88 percent of the towns and municipalities with only twenty being excluded. Two of the agencies, Housatonic and Northwestern, were not functioning to an extent that would permit the administration of the weighting instrument. However, the coverage offered by the thirteen remaining agencies appears sufficiently broad, in terms of both population and land mass, to result in an accurate testing of the proposed hypotheses.

Secondly, the regional approach, operating through an existing organizational structure, conformed to the projects' time and money limitations. Manpower constraints, combined with the statistical requirements of random sampling, prohibited the canvassing of the state and the development of a set of relative weights by use of a nonselective sampling procedure. In confining the investigation of public attitudes to the members of regional planning commissions, supplemented at times by leading area figures who were asked to participate in the weighting process, the results were opened up to several of the criticisms and limitations outlined in section 4.1. These limitations focus upon the strength of the link that exists between the tabulated weights obtained from the selected respondents and the true values of the community as a whole. In recognizing this point, the results of the weighting process must be closely compared to the characteristics of the region in order to determine their accuracy and to point out both the source and extent of potential variation.

Finally, the use of a regional approach allows the results to extend beyond the purely local problems which might arise as particular neighborhood impacts are evaluated. This approach has the obvious advantage that the respondent is more likely to react by providing a metropolitan set of responses as the result of his present participation in a regional planning body. This cosmopolitan attitude is helpful, since in a democratic society decisions should be ideally made on the basis of the broad majority point of view. It is true that, by asking commission members to reply to the weighting instrument on the basis of their own town's relationship with the potential highway, an element of parochialism is introduced. However, even here, there is still a tendency to reply to the question with the general good of the community in mind.

In approaching the problem of developing a set of relative weights, the goal was to derive a set of state-wide trade-offs that could be compared to the more local values in order to point out any potential areas of conflict. Considering the information revealed in the survey of the literature, it is clear that this is the first attempt at developing a state-wide highway investment profile as it applies to the nonuser impacts of road construction. If this goal were to be achieved, it was obvious that the approach taken must divide the state into manageable segments. The regional format served this purpose quite well as an initial starting point. However, in future investigations, where the actual selection of a corridor is to be made from a group of potential routes, it may still be preferable to use a finer subdivision within the region as an index of the nature and extent of localized concerns. In assessing the purpose for which the project was originally undertaken, however, it was felt that the regional approach, which would allow for aggregation up to the state level, would more precisely fulfill the present needs of the Connecticut State highway planners.

In offering the results obtained from the weighting as a partial indication of the relative importance of these items throughout the state, it is essential that the composition of the individual planning agencies be diagnosed and offered for public review. The members of the planning commission are usually drawn from the general public and, in most cases, are either business or community leaders from the town they are representing. Each town on the commission has a minimum of two representatives, one of which must be from or appointed by the town planning commission if such an agency exists. The other representative, or both in some cases where a town planning commission does not exist, are to be appointed as specified in the local ordinance. The usual procedure outlined by the towns involves either appointment by the chief town official, appointment by the governing body of the town, or direct election by the population. In addition to the two required representatives every town with a population of over 25,000 is allowed one additional representative for each additional 50,000 residents or part thereof.

The normal result of this representation procedure is that a suburban bias is usually attached to the agency. This bias, moreover, is predominantly toward the white property-owning middle class with a business orientation in those regions which possess a recognizable core city. In recognizing this bias, a correction factor was introduced into the sample. This correction factor relied heavily upon the professional judgment of the planning director, by allowing him to invite participants from outside the commission to assist in the weighting process. The additional participants were usually from recognized social or community action groups located in the region. On other occasions, religious representatives were asked to participate, as well as some newspaper people who were familiar with the character of the area. For the most part, these additional participants represented the urban core and tended to balance the opinions of the planning group with minority representation. The nonmembers only once exceeded 25 percent of the participating commission members responding, and the planning directors were impressed with the fact that a reversal of the bias in the opposite direction could be introduced if heavy emphasis were placed upon minority representation. It was felt that this reminder, and the desire on his part to develop a true picture of the region's attitude toward highway planning, would tend to guide the director's selections.

As a result of the above correction factor, it must be pointed out that the weights presented in Chapter 5 and summarized in Chapter 6 do not necessarily reflect the views of the planning agencies, nor are these results endorsed by them. The agencies retain the freedom to accept or reject the conclusions reached and to comment on them in any way they feel appropriate. This must be strongly emphasized, furthermore, since on only a few occasions were all the members of a given commission able to attend the meeting at the time the questionnaire was administered. The purpose of the regional approach was to facilitate contact with leading members of the community who would consider the impact of highway construction from a regional point of view. The planning agencies provided a vehicle through which such contact could be made on a group basis and, therefore, the responses obtained should be viewed with this limited purpose in mind.

In addition to the members of the planning commission, weights were also tabulated for two identifiable supplementary groups. The first was the individual planning directors. Obviously, their cooperation was necessary from the opening phases of the operation. Consequently, in order to familiarize them with the material to be given to the members of their commission, it was necessary to first administer the weighting instrument to the planners as a group. The results of this phase of the operation were tabulated and presented to the planners to facilitate the comparison of their particular answer with the views of their planning commission. For the sake of anonymity, the responses of the individual planners are not published. However, their answers are aggregated on the basis of population for each region. The results express the state profile for the twelve participating planners regarding the relative importance of the nonuser or community impact of highways.

Obviously, the regions previously excluded from the tabulation are also excluded from the results in the planners' column. Also, due to unforeseen difficulties, the responses obtained from the planning director in the Valley Regional

Planning area are also omitted. The use of this planning group allows the comparison of items, weights, and rankings between an aggregate group of concerned citizens and a professionally trained group of community developers. This comparison should facilitate an investigation of several of the hypotheses offered in the previous section regarding the pattern of responses to be obtained from each of the groups. In particular, an element of detached "objectivity" may be detected. Such a pattern would allow the planner to vary his responses from those of his planning group, and point the planners more in the direction of recognizing the need for solution of current regional problems. The results may also support the contention that planners are more concerned with items which bear fruit in the future and those public goods which require the recognition of the needs of the community as a whole. Finally, if the responses of the planners tend to differ from those of their planning commissions it may be that these differences tend to support the minority points of view that exist within the community. These views may lack adequate representation on the commission, or the persons impacted may not be fully capable of formulating and expressing their attitudes toward the affects of highway construction.

The second supplemental group was composed of key members in the Connecticut Bureau of Highways. This group was selected by the director of research and the director of planning working jointly to convene a group of highway officials who are active in various stages of the planning and decision-making process. The men participating in this phase of the operation ranged from those in charge of the earliest phases of corridor route location planning to those responsible for the design, scheduling, and land acquisition aspects of actual highway construction. The answers supplied by this group yields a standard for comparison with both the aggregate of answers supplied by the planning commissions and by the individual planners. Although the latter two groups answered the questionnaire with a particular region or town in mind, the highway officials were forced to reply to the weighting instrument based upon their average attitude toward these items within the state as a whole. Consequently, there is some room here for honest differences of opinion on the aggregate or state level as well as the normally anticipated variation between state and regional attitudes towards particular items. Clearly, a comparison of this type points out the sources of potential controversy regarding corridor site locations and, at the same time, it formulizes the view regarding the actual trade-offs used by highway planners. These trade-offs are currently applied in route location decisions and their formal expression will allow for further analysis and explanation where these answers appear to differ significantly.

4.3 The Format and Administration of the Questionnaire

Format

The questionnaire is a presentation of the thirty potential impacts developed in section 3.9. These impacts are divided into six categories, enabling the items to be reviewed according to dominant characteristics. The placing of five items in each category is done for both the simplicity of tabulation and presentation. The arrangement is at best experimental, and only suggestive of one alternative means by which the items can be organized for comparison. To some, the classification of items may be arbitrary. Consequently, the future application of the questionnaire to a specific corridor proposal may necessitate a shifting of items between categories or the creation of new categories to handle additional items which are unique to the route under consideration. Each of the items is provided with a description which translates the definitions outlined in section 3.10 into terms which are meaningful to the respondent. These descriptions provide additional information, which is intended to supplement the respondent's existing knowledge concerning the potential effect of these impacts upon his region.

The items were arranged according to a table of random numbers which allowed for the elimination of bias attached to the positions assigned each potential impact. The same random selection process was used to arrange the order in which the six categories themselves appeared. After arrangement, the items were presented twice, first with the assumption of a hypothetical detrimental impact upon the community and then, second, assuming the impact is beneficial. This dual presentation is a recognition of two elements which guided the estimation of community attitudes. The first is that local circumstances surrounding the selection process are very influential upon the actual rating assigned. Such items as the current state of employment conditions, the extent of the local tax burden, or the quality of social linkages can have significant effects upon the importance of the impacts as they result from highway construction. Secondly, aside from surrounding conditions, it is possible that the respondents may feel quite differently regarding the value attached to a beneficial as opposed to a detrimental effect upon a given item. This lack of symmetry may be a true and lasting expression of the individual's level of concern that he attaches to the particular item.

In the instruction sheet which accompanied the questionnaire, the participant was provided with a hypothetical highway construction example that he was to use as the frame of reference for performing the weighting function. The presentation of the example was necessary for two reasons. First, it gave each of the participants the same starting point for the weighting process. Since the goal of the questionnaire was to provide the Bureau of Highways with a generalized set of weights which could be used as a guide in the corridor route location process, it was necessary to frame the responses within a general format. Obviously, since no given project, either past or present, was sufficiently large to affect all of the regions throughout the state, it became necessary to refer to highway construction in the abstract. Secondly, the general example was designed to place the individual respondent on his own margin of concern regarding the importance he attached to each of the hypothetical impacts. The example was constructed so that all of the items were of concern to the respondent, with the least of them having just achieved a level of importance that would bring it to the attention of the person doing the weighting. The set of responses would then indicate the order in which the importance of these items appeared, and their intensity of concern, as indicated on the physical scale, would indicate the relative trade-off ratios involved. The example was worded to achieve this result without eliciting an over reaction on the part of the respondent. Care was taken not to lead the respondent by the structure of the example, nor to color his replies by the use of emotional statements or descriptions of the potential impacts.

The location of the example was assumed to be within the town of residence or within the "community," whichever was more meaningful to the participant. Also, for the sake of generalization, neither the nature of the project nor its size were specified within the example. In wording the statement in this way, it forced the respondent to rely primarily upon his own information system to supply the details associated with the potential construction of highways in his particular town. Furthermore, in connection with this protection of the general nature of the results, none of the thirty impact items specifically quantified the size of the anticipated effect. Obviously, if numerical entries were provided, which detailed the size of the physical impact, the rankings would be associated with these numerical coefficients and any alteration in impact size might lead to a change in the answers obtained. Such a quantification would be undertaken in the future where specific route location decisions are actually contemplated. However, for this initial phase of the operation, such quantification was deemed unnecessary.

There were three supplemental sections attached to the questionnaire. The first contained biographical material which was aimed at revealing the characteristics associated with the group of participating respondents. It was hoped that the correlation of the personal characteristics with the pattern of responses would help to explain any answers which could not be accounted for by the general nature of the region itself. The second supplemental section contained the first open-ended or general response question, which was included for two reasons. First, it was necessary to start the respondents thinking in terms of highways and highway impacts. This is especially true where it can be anticipated that a project will have beneficial as well as detrimental effects upon the community. In order to achieve this frame of mind, it was felt that, without any other instructions, the respondents should be asked to jot down the items that they would associate with the hypothetical transportation example as it is constructed within their community. The second goal in this section was to develop a set of initial concerns whose importance could then be compared to the ranking assigned to the item within the relative weighting scheme. Such a comparison might help to indicate whether or not the initial concern was of a topical nature which tended to disappear when the totality of potential effects must be compared in terms of their relative importance. It was hypothesized that the items which appeared in this section would receive somewhat less attention in the weighting process when the respondent was forced to consider all of the impacts.

The final supplemental section contained a second open-ended question which asked for any comments that the respondent might have regarding the questionnaire which he had just completed. It was hoped that three sets of comments would be obtained. First, even with the extensive pretesting of the questionnaire, there was a strong concern that there may have been important items which were omitted from the list of thirty. As a result, where these items were apparent to the respondents, it was hoped that they would supply information regarding them which could then be used for further reference purposes. Secondly, there was some concern regarding the format of the questionnaire itself and the mechanical problems that the method of weighting might present to the respondent. Primary interest was directed toward comments which referred to the level of difficulty associated with the operation and those which indicated that the participants were not physically able to perform the relative evaluation of the items. Finally, there was also some concern over the respondent's personal attitude toward the validity of his answers. It was important that, in performing the physical operation, the respondent should express his true feelings toward the value of highway construction, and not reply solely to the instrument placed before him. The provision of a second open-ended question gave the participant the opportunity to comment on these three items and to convey to the interviewer any other personal feelings that he might see as relevant to the analysis or tabulation of the results.

In developing the format of the questionnaire, two alternative approaches were considered. The first involved the presentation of a series of statements which asked the respondent to make a value judgment regarding the relative importance of the impacts as they were compared two at a time. In making his reply the respondent would be given the opportunity, in the extreme, to state whether he strongly agreed or strongly disagreed with the statement as presented to him. Other categories would be provided which would account for levels of concern that were located between the two extremes. This alternative was rejected for the following reasons. First, the number of comparisons which would have to be made to allow for the development of the complete set of relative weights would be so large as to be unmanageable. The multiple combinations of thirty items taken two at a time would require a questionnaire whose physical operation would be beyond the endurance of most of the participants. Secondly, the results of such an approach might fail to give a set of numerical weights which could be applied within the framework of the matrix developed in section 3.5. Since general weights are essential to the full testing of the hypotheses, it was felt that some alternative format must be adapted. Third, any numerical quantification, either implied or expressed, relating to the impact items, would again establish a set of trade-offs which were only valid for the specific example cited. As noted earlier, if the size of the physical impact were altered, this might very well affect the level of concern and, consequently, the response to the statement.

The second alternative involved the construction of a questionnaire which asked the participant to compare all thirty items simultaneously, rather than splitting them into smaller groups. In this approach, the items would be presented one after the other with the respondent then performing the appropriate operation. Although this approach applied part of the mechanical operation actually used in the questionnaire, it was felt that the necessity of operating upon thirty items at a time would be too confusing to the participant and would only tend to irritate or tire him. Obviously, the validity of the responses would be brought into question if the participant became bored with the operation as he moved towards the bottom of the list. A diminished level of attention would tend to make him careless regarding the weights attached to the item under consideration. Also, by presenting thirty items simultaneously there was a greater probability that the respondent would make his answers to the latter items a function of his knowledge regarding the previous items. In some cases this might lead to confusion or the desire to stack the responses unfairly in the direction of one or more of the impacts.

The Mechanical Aspects of Replying to the Questionnaire

In evaluating the relative importance of the items contained in each of the eight categories, two points were impressed upon the respondents. First, it was made perfectly clear that they were to use the transport example as their frame of reference. Moreover, the participants were to assume that all of the impacts listed in each of the categories would occur. This latter assumption was emphasized in order to remove any doubt that a participant might have regarding the intent of the items. Specifically, he was not to consider the probability that each of the items was to occur, but rather he was to base his reply on the assumption that each of the impacts will definitely occur.

The second point was that the respondent should assign his weights with his own town in mind. In addition, he should assume the role that he now plays as a member of this planning commission representing his town in providing the answers. The purpose of this latter instruction was to provide the respondent with a known framework for reply. This would tend to eliminate or reduce any confusion which might arise regarding the point of view he was to assume or the point of reference for the operation. Specifically, he was not to guess at what the people in his town would do in reply to this questionnaire. This would remove the need for the respondent to move outside of his normal response pattern in supplying the required answers.

The items were considered in eight separate groups and the respondent was asked to rank the factors in terms of the order of their concern and then to rate them on a physical linear scale running from 0 to 10 according to the importance that they felt each deserved in the route location process. The two-step operation was a simplification of the weighting process. The ranking was done first in order to organize the items in terms of their relative positions. Then, after this task was accomplished, the ranking was used as a guide for the rating on the physical scale. Here an item of maximum importance was assigned a rating of 10 while an item of no importance was assigned a rating of 0. A continuous scale was presented between these two extremes in order to properly register the precise intensity of concern. This operation was performed for both the potential beneficial and detrimental impacts of each of the items.

Groups 2 through 7 required a comparison of the thirty items as organized according to their dominant characteristic. Group 1 required that the participant rank and rate the relative importance attached to the user as opposed to the nonuser or community effects of the highway project. The last category, group 8, asked the participant to assess the relative importance that he attached to each of the six major categories under which the impact items were located. The purpose of this operation was to allow the respondent to make a judgment on the relative importance of the entire collection of items contained within the categories. If the operation was correctly performed, this would allow the investigator to assign weights and compare trade-offs across all thirty items.

The Administration Procedure Accompanying the Questionnaire

As indicated earlier, the questionnaire was administered to the assembled members of the planning agency supplemented by a limited number of respondents outside the commission. The number of participants per region ranged from a low of 7 to a high of 22 with a total of 226 valid responses. This figure includes those answers supplied by the 12 planning directors and the 13 participating highway officials. Both the instructions and the questionnaire itself were read aloud at all times. This was done in order to fully control the administration of the questionnaire and to provide the respondents with a pacing mechanism which would indicate the length of time they would be able to devote to each section of the instrument.

In almost 90 percent of the cases, the respondents were able to complete each section of the questionnaire in the time allotted. However, approximately 10 percent of each participating regional group was unable to perform the required operation within the allotted time span and, as a result, this number served as the discard rate for the improperly answered questionnaires. This failure held steady throughout all of the thirteen regions with the highway officials and the group of planners containing no improperly answered questionnaires. The reading of the instruction sheet and the questionnaire was also undertaken in order to facilitate the answering of any questions which might be asked by the participants regarding the wording or intent associated with a particular item. To keep the same question from being repeated, and therefore disturbing the group as a whole, it was felt that all participants should be at the same point in the answering operation in order that all might view the item in question.

The administration time took slightly less than one hour in most of the regions with the longest period of testing being approximately one hour and twenty minutes. Within the average time span the administration of the instructions lasted slightly under fifteen minutes, while the actual process of responding to the instrument took slightly less than forty-five minutes. This latter operation was facilitated by the respondents seeing the same items repeated for a second time on the beneficial side.

In verbally administering the questionnaire, the same interviewer was used in all cases except one. This single exception existed in the Southeastern Regional Planning area, with the tabulated results showing no apparent affect from the change in the administration pattern. During the period of administration, the participant was continually reminded that he must answer the questions with respect to the general transportation example. Moreover, it was also indicated at several points that he was to assume that each of the impacts were to definitely occur within his town of residence. The respondents were also asked to perform the ranking function first and then to use this as a guide in establishing the physical intensity associated with each particular impact. To facilitate this operation, after each of the categories and its items were read aloud, the interviewer issued the following verbal instructions:

If all of these impacts will occur in your community as the result of the construction of the highway, which ones should be considered most important in the planning and route location process? And which ones would you want to have more information about before you passed judgment on this project?

This quote was repeated at every phase of the operation so that the participant had no doubt concerning either the location of the impact or its probability of occurrence.

In general, the majority reaction to the questionnaire and its verbal administration was favorable. Initially, one or more participants in each of the groups tested expressed a degree of hesitation regarding either the validity of this approach or their ability to think in terms of relative trade-offs regarding the impacts of highway construction. However, in only one instance was the administrator unable to resolve the difficulty. In this case the hesitation progressed to the point where the participant did not participate in the questionnaire.

During the postquestionnaire discussion period, the feeling among most of the participants was that the instrument fairly represented the alternative impacts associated with highway construction. In fact many were surprised that so many impacts were to be considered within the investment decision-making process. This favorable reaction existed to the extent that several of the respondents felt the questionnaire itself was highly informative and educated them personally to the need for trade-offs with respect to highway construction. Finally, several of the participants acknowledged that they could easily see the application of these items within their own community, and they felt that particpating in this particular interview had altered somewhat their attitudes toward both highway construction and the quality of the planning process.

4.4 Formalization of the Operational Hypotheses and the Process of Verification

The statements outlined in this section are operational hypotheses whose validity can be tested by appeal to the tabulated results obtained from the questionnaire. In general, an hypothesis is a statement regarding the nature of the real world. It describes one or more aspects of the problem under review and draws a relationship between the proposed solution to the problem and the appearance of the problem in a practical setting.

In applying the term hypothesis to this particular work, there has been a distinction drawn between those hypotheses which are purely *structural* and those which are *operational*. The structural hypotheses, outlined in section 3.5, related to the abstractions which are necessary to develop the matrix approach to project evaluation. These abstractions simplified the problem and provided a starting point from which the arguments surrounding the matrix approach were structured. On the other hand, the operational hypotheses, which are listed in this section, are statements that attempt to organize the observed facts and explain them in terms of the underlying behavioral or technical relationships. The term technical relationships refers to the affects that the physical laws of nature have on the observed results. These relationships are not of prime concern in this particular evaluation process; however, behavioral relationships are relevant and refer to the patterns of response associated with human characteristics or activity traits. In this particular work, the behavioral relationships possessed by the participants will hopefully be related to the stated hypotheses through the tabulated results obtained from the questionnaire. If this relationship is substantiated and, as a result, the validity of the hypotheses are supported, then the results of the investigation will allow an evaluator to make predictions regarding future events based upon his knowledge of behavior patterns. The essence of this substantiation process rests, therefore, upon the verification of the hypotheses. This verification in turn depends upon the weight of the evidence presented. Consequently, in establishing the validity of the operational hypotheses, the evidence obtained from the questionnaire and its quality become a key feature of the investigative process.

In judging the quality of available evidence and the extent of support that it lends to the establishment of the proposed hypotheses, there are two generally accepted methods of verification. The first is statistical and is generally acknowledged as being objective, since it relies upon the techniques of probability and inference as the basis for justification or acceptance. Either through the correlation of the evidence with past events or an evaluation of the accuracy of future predictions, the statistical method can offer support for the observed relationships within specified ranges of probability. Such an evaluation process allows the investigator not only to estimate the extent to which the observed variables have been related to each other in the past, but it also allows him to establish a probability regarding the possible accuracy of this observed relationship.

The second method of verification is subjective and relies upon the testimony of recognized authorities, who, based upon their knowledge of the area, offer verbal support as to the quality of the observed relationship. In particular, this testimony comments upon whether the observed facts are consistent with the characteristics as known by the expert. This form of support is, in many cases, less impressive than the mathematical precision introduced by the objective statistical approach. However, in establishing the validity of a proposition or the relationship between observed facts and possible explanations, the comments offered by experts in the field are useful in developing the weight of evidence in support of a particular position.

If the testimony of experts is relied upon in part to establish the quality of the evidence and its relationship to the hypotheses, then two subissues are necessarily introduced. The first involves the quality of the credentials possessed by the expert, which helps to establish the validity of his response. The second is the ability of the witness to respond objectively and to pass judgment on the issues with a proper amount of incite. This combination of sensitivity in observation and professional skill in evaluation is essential. In this particular case, the evidence to be reviewed is the tabulated results of the questionnaire obtained from each of the thirteen planning regions, while the expert testimony of a witness trained in the area is to be obtained from the regional planning director. His evaluation of the results and comments regarding their relationship to the behavioral characteristics of the respondents will be relied upon as a test of the operational hypotheses.

The following operational hypotheses refer to the anticipated responses from the questionnaire. They are offered as an indication that the technical evaluation process, outlined in sections 3.6 and 3.7, is fully operational with respect to the establishment and application of the relative weighting of impacts. The hypotheses are:

1. That representatives of the general public can perform the physical operation required by the questionnaire and establish the relative impact weights
2. That individuals, acting as representative of the community, with the proper qualification and guidance, are both articulate and cognizant enough of their own values to be able to describe these values in a way which can be

applied to corridor route location decisions. In establishing relative impact weights based upon these values, it is recognized that these will tend to change overtime with alterations in both the individuals information system and the circumstances which surround a particular corridor proposal. As a result, any statement of impact weights must be periodically updated in order to account for these fluctuations.

3. That significant variations in community values and weights will exist between geographically identifiable regions. These variations are sufficiently strong so that a properly developed test instrument can detect them in a way which is meaningful for highway investment decision making. This hypothesis, if substantiated, would lend support to the corollary that such variations also exist between localities within the test region
4. That area residents possess the potential to feel differently regarding the beneficial as opposed to the detrimental impacts of a highway as they might affect the same factor. This potential will manifest itself in a different relative ranking for a beneficial or a harmful affect of a given factor on the community. Such variations in ranking are strongly influenced by the present conditions of the social and physical environment which exists at the time the corridor decision is being made
5. That the weights as described by the participants can be viewed by impartial persons trained in the area, who can comment authoritatively on the validity of the results and provide verbal explanations relating them to regional characteristics that will yield an in-depth profile of the community value structure as it relates to highway construction

In offering the above hypotheses, there are certain limitations introduced by the use of a questionnaire approach. First, in this initial test of the plausibility of the matrix process of investigation, the questionnaire was structured in a general way which would allow it to be applied in each of the thirteen operating planning regions in the state. As a result, neither the size nor the location of the route were specified leaving these pieces of information to be filled in by the respondents own information system. Further the size of the potential impact was not quantified numerically in order to preserve the generality of the observed weights. Obviously, in cases where actual corridors are being considered, these three items would be specified with the resulting weights being applicable to the given problem under consideration. In our test case, however, we assumed that the participants could abstract from the transportation example which was provided and that they were sufficiently knowledgeable regarding the nature of their town to relate the potential impacts of the highway to the underlying characteristics of their area.

A second limitation is introduced by the written descriptions associated with the potential impacts. Two concerns appear with respect to this item: First, that the wording of the statement might be emotional or inflammatory, leading to a biased response on the part of the participant. Regarding this item, care was taken to avoid describing particular impacts in emotional terms while, at the same time, an effort was made to avoid obscuring the issue to the extent that the respondent would not be able to understand the true meaning of the item. The second concern was that the importance of the questionnaire might not be understood at all by the participants. It is necessary that the participant have a sufficient understanding of both the purpose of the test instrument and of each of the statements to react intelligently and in a manner which would allow for his presentation of his true feelings towards these items. Obviously, there was an interest in having the respondents react uniformly towards these statements with a minimization of perverse reactions towards the impacts listed. Such perversities could arise where respondents felt that the barrier effect of a highway was actually beneficial to the community and therefore desirable, or where the elimination of business firms from the region was advantageous due to the possibility of improving the quality of the environment. Responses such as these were not anticipated by the questionnaire and would have rendered the results meaningless if they had appeared repeatedly in the weighting process.

The third limitation is that the participants may not have previously thought in terms of the multiplicity of impacts which can arise from highway construction. As a result, when presented with these items for the first time, they might think it difficult to react in terms of trade-offs. Consequently, the responses are directed only towards the mechanical operations required by the questionnaire and do not reflect the true feeling of either the respondent or the community as a whole. Accompanying this problem is the difficulty that the participants might not be able to think in terms of individual impacts. They might only see the interactions that exist and would not be able to separate out actual or perceived lines of clausality. In recognizing this limitation it was felt that such problems would manifest themselves in the pattern of response obtained from the questionnaire. Where this problem was identified, the proper corrective action could be initiated.

A fourth limitation concerns the representatives of the group which is selected to perform the weighting process. This particular problem has been considered at length in sections 4.1 and 4.2 and will not be reviewed in detail here. In reply to this point it must be made clear that the questionnaire, as presently constructed, exists solely as an initial test of the validity of the overall weighting approach. Its purpose is to provide an indication of the potential impact items which are of true importance to the community and to determine whether such importance can be established through community participation. As a result the complete accuracy of the results is not as essential as the establishment of the principle that the approach itself is valid. While the composition of the test group partially governs this validity, there is an acceptable margin for variation between the answers obtained and the actual values held by the entire community.

The fifth and final limitation concerns items which relate to the administration of the test instrument. Specifically, the respondents may have felt that since the items were repeated twice that there should be an element of consistency introduced into both the weights and ranks assigned the item on both sides of the ledger. Although it was pointed out that consistency was not to be considered a virtue in this case, the respondents may have felt that there was a psychological deception being applied here and, consequently, they might have tried to out-think the research team by providing the same order and magnitude of answers on both the beneficial and detrimental sides. The difficulty in recognizing the actual appearance of this problem is that plausible explanations can be developed for both consistent and inconsistent answers. Consequently, there is no complete test which will provide an estimate of the extent to which this limitation has actually occurred. As a result the research team can only assume that the participants accepted the instructions at face value and that they did not attempt to anticipate any desired pattern of response.

Another administrative problem involves both the length of time needed to complete the answers and the general repetitive nature of the operation and the items. It is possible that the time needed to administer the questionnaire might have created boredom in the participants so that the quality of their replies tended to diminish as the questionnaire wore on. Also, their familiarity with the mechanical operation and, in the second part, with the terms themselves may have led to a carelessness in the way the responses were made. In both of these cases the replies would be invalid and reflect a response solely to the instrument rather than a thoughtful application of community attitudes towards highway construction. The only way to detect this problem would be by engaging in post questionnaire discussion periods which might reveal the participant's attitude toward the instrument and his general knowledge regarding the impacts of highway construction. In these discussions the instances of overfamiliarity and boredom might be revealed or at least suspected.

In testing the validity of the responses two possibilities were outlined earlier in this section. The first relied upon the objective test based on statistical analysis where the results are numerically tested in light of the hypotheses, and coefficients are derived which indicate the extent of accuracy. Unfortunately, the elements necessary for such an objective testing are not present in connection with the tabulated results of the questionnaire. Some index of the validity of the hypothesis would have to be present as a dependent variable. This would allow for its comparison to a set of independent variables that represent behavioral characteristics of the population and which can be applied in an explanatory manner. Neither of these two sets of data are present. Consequently, the objective test of validity must defer to the subjective evaluation based upon the testimony of experts in the field.

The method chosen for this evaluation is a modification of the Delphi Technique, developed and tested by the Rand Corporation for the United States Air

Force.[4] In its pure form, the Delphi Technique is a formal procedure for deriving an opinion consensus from a group of respondents. In its initial stages, the participants, through an administered questionnaire, individually supply answers in numerical form regarding the variables under consideration. The results are tabulated and resubmitted to the participants in a controlled feedback situation, on an individual basis, to allow them to see their results compared to those obtained from the group as a whole. They are then asked to either justify their conclusions or make corrections where they feel necessary. These individual corrections are then tabulated and the results fed back a second time, again on an individual basis in order to elicit further reappraisals of the quality of the average result as opposed to the comments made by the individual. This process of reappraisal is continued until the administrator is satisfied that as much of a group consensus has been achieved as is possible given the recognition that valid differences do appear regarding opinions on various items.

The advantages of this approach are threefold. First, it provides anonymity to the respondents by having their replies to the questionnaire and to the reappraisals obtained in private. Consequently, their answers are not influenced by a dominant participant as much as they might be if the administration were undertaken in a group. Second, the process of controlled feedback allows the administrator to eliminate unnecessary items and to focus upon those issues which point out major conflicts or which are of prime interest regarding the problem at hand. Third, once achieved, the tabulated results do provide a statistical index of the average opinion supplied by the members of the group in the aggregate. Such a tabulation allows for a quantitative review of the importance associated with each of the items involved.

As developed in its modified form for this particular project, the Delphi Technique has been adapted to the problem of establishing relative weights given the time and money limitations which are introduced by the size of the study. The first alteration took place in the method whereby the questionnaire was administered in the first round. Recognizing the fact that individual interviews, which would allow for an overview of the state as a whole, could not be performed in a reasonably efficient manner, the test instrument was administered on a group basis. The administrator attempted to minimize the nontechnical questions that were raised from the floor, and which might bias or tend to sway the responses made by the other participants. In the reappraisal round the mean responses were resubmitted to the individual planning director on a private basis showing him both the responses to the commission members and his own personal replies to the test instrument. At this time the planning director's appraisal of the results were obtained and, where he felt sufficiently satisfied, he was asked to provide a further analysis of the regional characteristics which might have accounted for the observed pattern of response. In returning to the planner, three specific question areas were posed to him:

1. How valid to you feel the relative rankings are for your region as a whole? Do you feel that the order in which the items appear gives a fairly accurate picture of the relative importance of each of the items?
2. What characteristics of the region might account for the importance attached to the items on both the detrimental and beneficial sides?
3. Do you feel that the weights as presented would alter as the town of resident of the respondent moved from the core city outward toward suburban or rural communities in your region? If these weights and rankings might vary, to what extent will this variation occur and what can be cited as the cause of this variation?

The use of the planning director as the reappraiser was based upon both their professional qualifications in the field or urban and regional analysis and their familiarity with the characteristics of the region and the composition of the planning commission. Specifically, the planners are regarded as the type of expert who could be called upon in actual future corridor location decisions to work with members of the Bureau of Highways and to interpret the results of both the identification and weighting phases of the matrix application.

The chosen technique is considered as a modification of the pure Delphi approach for the following reasons. First, it strives for a consensus view based upon a reappraisal of the qualities of the tabulated results. In accepting this need for review, the approach recognizes the human element involved in corridor location decisions and the effect that this process can have on both the social and physical environment of the community. Secondly the controlled feedback of responses focuses upon the issues of major concern which might appear in individual cases. In doing this, it allows for the reduction of tension or resolution of conflict in those areas where wide divergence might exist between the weighted impacts assigned by the members of the community and those assigned by either the highway officials or the state as a whole. Finally, to the extent that the questionnaire is administered on an individual basis or to a group with a minimum of comment being allowed, the anonymous nature of the replies are maintained. Not only will individuals be free of social pressure at the time the answers are supplied, but their personal pattern of response will be suppressed by the aggregation of results to achieve a mean set of replies for the community as a whole.

4.5 Method of Tabulation

The previous three sections have centered upon the technical procedure for relating the value structure of the community to the effects of highway construction. The questionnaire technique, as outlined there, relies upon public participation to perform a relative weighting operation that yields a set of information relevant to the selection of the most desirable highway corridor. This section provides a description of the statistical technique that was used to bridge the gap between the personal act of transforming abstract individual values into numerical or distance dimensions, and the resulting weights which are to be applied within the matrix framework.

The weights can be tabulated in a number of ways that emphasize various aspects of the response pattern and that yield differing degrees of information. The method selected appears to present the weights in their most useful form and to eliminate undue distortions, which can deceive the reviewing body. Other methods are available, however, and the interested reader may consult the listed supplementary sources for a more extensive review of all possibilities.[5]

The tabulation procedure adopted the following technical format and relied solely upon data supplied by the respondent in the linear rating operation. This operation was performed on the right-hand side of the questionnaire. The line that appears there is exactly ten centimeters in length, with each unit from 1-10 representing a uniformly increasing level of personal concern for a particular item. The numerical rating (i.e. *raw score*) was transferred to computer coding by use of a key-punch operation. After these two steps were accomplished, the following tabulation procedure was applied.

Categories II-VII contained the thirty potential impact items arranged under six general *groups*. Each respondent had his answers arranged on four data cards in group-item order, which were read into the storage cells of the computer on the basis of each of the fifteen participating *units*. The decimal weights attached to each item by a given respondent were then calculated by summing the raw scores associated with all five items in a specific group and dividing this *total score* into the raw score allocated to each of the five items. The result is a weight showing the relative level of concern allocated to each item within the group.

At the completion of the above operation, each participant's item decimal weight within a group, is weighted a second time by multiplying the item weight by weight afforded the group as a whole in the Category VIII calculations. The weights of Category VIII were computed in the same individual manner as were those in Categories II-VII. The percentage weights per individual, which have been operated upon twice (once within a group and once among groups), are summed over the number of individuals participating within a unit. This total is then divided by the number of participants to yield the average relative percentage weight assigned within a unit by the aggregate of participants.

An example illustrating the above operations is as follows. Assume that there are two judges (J) who are asked to rate two separate groups (G) containing three items each (I) according to some given characteristic. The results of this rating might appear as shown in table 4-1.

Here the raw item scores are summed to yield the total raw score (TRS) used as the denominator in the next operation. The results of this second step are shown in table 4-2.

Table 4-1
Sample of Raw Score Results

Judge	G-One Item A	B	C	TRS
1	9	6	5	20
2	8	10	7	25
Judge	G-Two Item D	E	F	TRS
1	6	3	1	10
2	7	7	6	20

Table 4-2
Decimal Results of Initial Division Operation

Judge	G-One Item A	B	C
1	.45	.30	.25
2	.32	.40	.28
Judge	G-Two Item D	E	F
1	.60	.30	.10
2	.30	.35	.30

If in a separate operation the judges had been asked to also rate the groups as a whole according to the same characteristic, the results might appear as in table 4-3.

These decimal weights for each group are then used to weight the individual decimal results in table 4-2. These answers appear in table 4-4.

The results are then summed columnwise across judges with the subtotal divided by the number of judges as shown. These concluding decimal figures are the average item weights assigned by the collection of participating respondents.

This is the tabulation procedure followed in developing the relative weights for each of the fifteen participating groups. The results in Chapters 5 and 6 show the relative importance of each potential impact within the broad category of

Table 4-3
Raw Score and Divisional Operation Resulting from Group Rating

Judge	Group One	Group Two	Group TS
1	9	4	13
2	8	7	15

Judge	Group One	Group Two
1	.6923	.3076
2	.5333	.4666

Table 4-4
Combined Weighting by Item and Group

Judge	G-One Combined Weight A	B	C	
1	.3115	.2076	.1730	
2	.1706	.2133	.1493	
	.4921	.4209	.3223	
Divide by the number of judges.	.24605	.21045	.16115	average item weight

Judge	G-Two Combined Weight D	E	F	
1	.1846	.0922	.0308	
2	.1633	.1633	.1400	
	.3479	.2555	.1708	
Divide by the number of judges.	.17395	.12775	.0854	average item weight

community or nonuser effects. To allow for a full relative comparison of both the user and nonuser effects, a third level of weighting must be introduced using the information provided in Category I. This additional weighting procedure would be accomplished in the same manner as the previous two.

To simplify the understanding of the above procedure it is helpful to shorten the example by presenting it in notational form. The following set of equations contains an abbreviated description of the computational format.

1. $\sum_{i=1}^{x} IRS_{ij} = TRS_j \qquad j = j_o$

where:

IRS_j = Item raw score for judge (J)

TRS_j = Total raw score for judge (J)

x = The number of items

j_o = The given judge

2. $\frac{IRS_{ij}}{TRS_j} = IWG_{ij} \qquad i = 1,2, \ldots, x; \quad j = j_o$

where:

IWG_{ij} = Item weight within a group

3. $IWG_{ij} \cdot WG_j = TIIW_{ij} \qquad i = 1,2, \ldots, x; \quad j = j_o$

where:

WG_j = Weight of the group

$TIIW_{ij}$ = Total individual item weight

4. $\sum_{j=1}^{N} \frac{TIIW_{ij}}{N} = AIW \qquad i = i_o$

where:

AIW = Average item weight

N = The number of judges

i_o = The given item

4.6 Alternative Levels of Use for the Weights of the Matrix

As indicated throughout the matrix presentation in sections 3.6 and 3.7, the purpose of developing an accounting format is to provide the decision maker with as much usable information as he can possibly apply in the corridor selection process. The goal is to alert the decision maker to the importance of the anticipated impact of highway construction, and to allow him to base his decision upon both this knowledge and his skill regarding the professional aspects of road building.

In striving for this end, it is recognized that there are severe limitations in both the time and money expenditures which can be devoted to the corridor location process. Obviously, as with any project, there are avenues of investigation which must either be eliminated or scaled down in order to conform to project limitations. Any matrix which is to serve as a tool of investigation must recognize this limitation and be flexible enough to adapt to both the needs and abilities of the planning agencies. To achieve its goal, the matrix must be capable of suboptimizing the output of the research effort given the above constraints. In other words, it must be able to achieve either a fixed informational output at least cost, or it must be able to start with a given input limitation and achieve the greatest amount of information relevant to the decision-making flow.

By developing alternative levels of application for the matrix weights, this desired extent of flexibility has been achieved. Moreover, through this effort, the nature of the matrix approach has been retained in the following ways. First, through the continued use of weights, the importance of the trade-offs, with respect to potential impacts, is maintained. Both the investigator and the reviewing public are reminded that none of the impacts anticipated from normal highway construction possess a veto power over the final corridor decision. Furthermore, in emphasizing the relative nature of various potential effects, the use of weights also impresses upon the planner the realization that he must make his decision on the basis of explicit trade-off values.

Second, the use of weights reinforces the need for some form of formal public participation in the corridor selection process. One method of increasing the amount of satisfaction on the part of the public with the ultimate decision is by allowing "public representatives" to participate in various stages in the decision-making flow. In assigning these representatives the task of establishing relative weights associated with the anticipated impacts, the participants are both indicating the communities attitudes toward the importance of each item and, in the process, educating themselves regarding the trade-offs that must be established in order to perform the decision-making activity.

Finally, the success with a partial application of the matrix approach may create the desire for an increased use of the technique in more complicated situations. The success of the approach on a limited scale may impress the local citizenry to the extent that those in a position to do so will request additional funds to carry out a more in-depth analysis, where the proposed project warrants a deeper consideration of the total impact.

There are three recognizable levels of application for the weights developed in the matrix. Each level, in ascending order, implies a greater degree of technical sophistication and research effort. In addition, the investigator himself must possess both a greater sensitivity to the community effects of highway building and either professional training or practical experience. These skills are required in order to allow him to anticipate and recognize the structural relationship existing within the community and those which will be created by the development of new road facilities. It is important to remember, however, that each of the three stages of investigation attempts to provide more information than was formally available under previous selection procedures, and to provide this information in an organized manner which will allow for its direct application to the decision-making process.

The first level of application involves the use of the existing weights on a regional basis to assist in the earliest stages of corridor planning. This stage involves the development and selection of various corridors which will serve as the alternative routes for impact evaluation. From this given set of corridors, one route will be selected which maximizes the net total beneficial impact upon the area. Up to now, the development of these alternative corridor proposals have been based upon three criterion of planning. First, they reflect desired lines of traffic flow which indicate points of origin and destination for trips generated by highway users. This consideration is based upon traffic service and attempts to minimize the overall user costs of travelling a given distance between two set points.

The second criterion involves the topography of the land within the desired line of travel. Corridors are established, where possible, in areas which minimize the technical difficulty encountered in physically constructing the proposed road. In general, the planner avoids encountering those areas where the terrain would severely hamper both the initial construction and safety of vehicle operation associated with the proposed highway. As a result, where possible, corridors would avoid mountainous land masses, poor soil conditions, potential flood plains, or areas which could be affected by seasonal storms.

Finally, the last criterion is based upon an informal knowledge and recognition of the impact that the corridor will have upon the community. Highway planners recognize the interrelationship that exists between road construction and the level of community activity, especially in terms of the economic and land use effects of a proposed project. As a result, corridors are now developed with an eye toward minimizing the undesirable aspects of economic or land use impacts upon the community. However, two important points are generally overlooked. First, as these corridors travel through different communities, the importance attached to each of these potential community impacts varies. Consequently, while some regions might desire further economic development, others may abhor the thought of road construction which brings greater industrialization in the area. Second, the planners may fail to develop their alternative corridor proposals with a view toward maximizing the potential community benefit to be derived from road construction. In many cases, the conscious development of the beneficial aspects of these proposed projects is neglected entirely. However, if properly constructed so as to aid the quality of the physical and social environment surrounding the network, resistance to new highway construction might be reduced.

With the introduction of regional weights which indicate the relative importance of potential impacts within the affected community, the informal nature of the route location process as it refers to community impacts is greatly improved upon. Corridor proposals can now be established for in-depth investigation based upon valid information regarding the way in which the potential impacts are valued within the region as a whole, at least on the general level. The weights provided in Chapters 5 and 6 are aimed specifically at this primary level of application. They indicate regional attitudes towards potential impacts on the abstract level, which should be immediately applicable, within limits of moderation, to the development of alternative corridors within the state. Obviously, if marginal changes in the intial level of corridor planning can knowingly minimize harmful effects or emphasize those impacts which are of significant value to the community, then the route has a greater change of acceptance. The set of weights provides the proper amount of information to enable this initial selection based upon the relative importance of the potential impacts.

The second level of application involves the development of a set of relative weights which apply directly to the project under consideration and which rate the importance attached to each of the identified physical impacts associated with a particular corridor proposal. This phase is clearly a further application of the matrix approach which stops short of the full quantification of the potential impacts in dollar form through the application of the quasi-market mechanism. At this level, all of the potential, both beneficial and detrimental to users and non-users, are identified and quantified within physical dimensions. These physical impacts are then assigned to the appropriate group as outlined in the analysis of the column elements in section 3.6. After the identification and assignment phases are completed, the physical impacts are then presented to the "representatives of the community" for weight assignment which would then indicate the relative importance of the impacts within the value structure of the community. This procedure would involve an application of a questionnaire similar to the one used to establish regional weights.

After the weights are tabulated, the highway planners can review the results and apply them as an indication of the importance attached to each of the physical impacts of the various corridor alternatives. Based upon this review and his professional skill and experience in highway planning, the appropriate decision maker can, at this point, choose between the alternative corridor proposals. Clearly, this decision is based upon more formal information than is presently included within the decision-making flow. However, there are difficulties introduced by the fact that the actual trade-offs used by the decision maker are still partially hidden from public review. Progress is made, however, in the area of explicitly focusing upon the trade-offs used in the planning process, by the introduction of a set of community weights which act as a standard of comparison in evaluating the decisions offered by the highway official. This standard can serve as a point of departure for public review, and should point out the areas of conflict between public and departmental assessment of the importance of each of the corridors. It also provides a basis for checking the actual postconstruction impact of the highway against the projected estimates upon which the decisions were based. Continued application of this checking process should improve the physical impact estimation process by revealing key structural relationships and thereby upgrade the quality of the decision-making flow.

Finally, the third level of matrix application involves the full development of the entire aggregation process which results in an index of relative value that enables the corridor decision to be made. This level is by far the most sophisticated and actually introduces a third step in the estimation process to accompany the establishment of weights and the quantification of physical effects. This third step in the estimation process involves the assignment of dollar equivalents to the physical effect upon the community. This process and its use in combination with the weights derived through community participation was outlined in section 3.7 and will not be repeated here. It is important to reemphasize, however, that this stage necessitates the greatest expenditure in both time and money, with the most information being available to the highway planner as well as the public. Such information presents a clear picture of the actual trade-offs used in the decision-making process and establishes a firm point of departure for review of the decision by a concerned citizenry. Clearly, by engaging in this sophisticated analysis, the total impact of a project upon the community can be afforded its proper relative position within the decision-making flow.

5 Analytical Profile of Three of the Thirteen Organized Planning Regions in Connecticut

5.1 The Selection of the Three Sample Regions

In order to verify the hypotheses established earlier, the thirteen organized planning regions in the state of Connecticut were utilized for field investigation. The weights for both the beneficial and detrimental impacts were established for each of the thirteen regions. Once these weights were developed, an intensive interview was held with each of the regional planning directors in order to further explore the nature of the weighting. Upon completion of this operation, a tentative "regional profile" was drawn up for each region and again submitted to the planning director for further comments. On this basis a final profile was developed.

It is not the intention of this chapter to look at each of the thirteen regions. Three regions have been abstracted for purposes of illustration and include an urban region, a semiurban region, and a rural region as shown in figure 5-1. The weights and profiles for these three regions are given in detail.

Chapter 6 shows the aggregate of all responses for the entire state in order to indicate regional variations, variations between professional planners and commission members, and between highway planners (who are engineering oriented) and the planners and commission members.

5.2 Analytical Profile of the South Central Connecticut Regional Planning Area—An Urban Example

The South Central Regional Planning Area is composed of the following fifteen towns: Bethany, Branford, East Haven, Guilford, Hamden, Madison, Meriden, Milford, New Haven, North Branford, North Haven, Orange, Wallingford, West Haven, and Woodbridge. The number of valid responses obtained in this region totaled twenty. In reviewing the pattern of responses, it appears that the participants were most concerned with the economic effects of highway construction modified by specific concerns with individual items in the health and safety category. Significant attention is also focused upon the land use category with the exception of historic sites and, to a lesser extent, housing. This set of answers seems to indicate that the residents of the area are highly concerned with the developmental aspects of highway construction with little or no emphasis placed

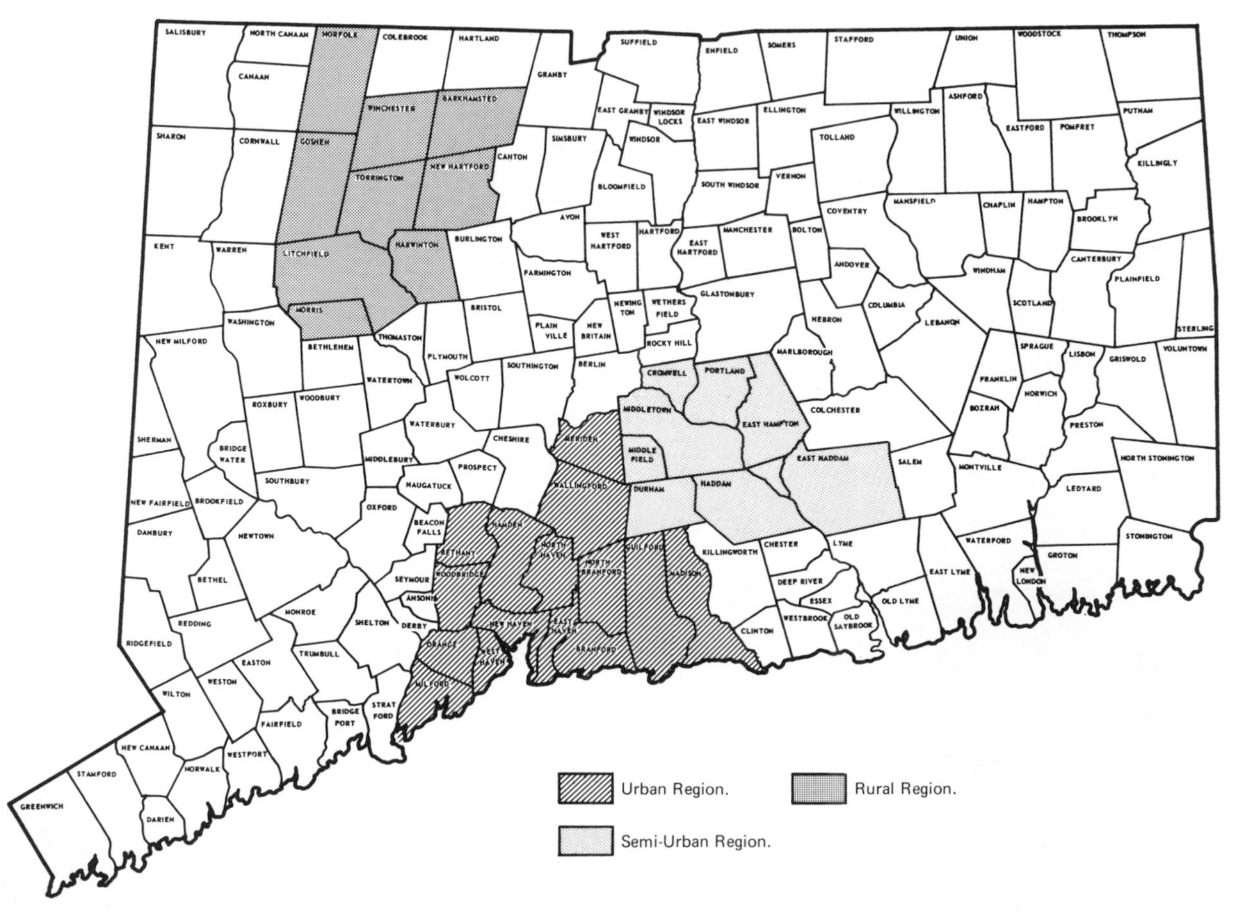

Figure 5–1. Map of Connecticut. (Regions Used for Illustrations Outlined)

upon the effect that a highway might have upon either governmental structure or the social character of the area traversed.

Extensive regional involvement with past highway construction should have given the participants a fairly broad background regarding the multitude of potential effects that highway construction can have. However, the differences that exist between the responses and the observed needs or characteristics of the area would tend to raise doubts regarding the complete accuracy of the results. Specifically, the profile of the region obtained from this ranking does not yield a full description of the area. This may be caused in part by three forces. The first is the result of the quality of past highway planning efforts in the south central area. In particular, there has been a high degree of cooperation and coordination of effort between the Planning Agency and the State Highway Department. This cooperation extends to the point that the agency, itself, participated extensively in the development of both the route and design characteristics associated with the construction of Interstate 91. This exceptionally close degree of interagency contact led to the construction of a route which provided maximum net benefit to the area, while at the same time yielding superior access for the north/south traffic flow. As a result, the harmful effects upon the region were minimized, while the beneficial effects were heightened and allocated in a highly favorable manner.

This degree of cooperation was unique to both the region and this particular highway. Moreover, it required special legislative attention to achieve its maximum fulfillment. The relationship of this planning activity to the responses of the agency members is simply that careful past planning has avoided many of the harmful political or special effects associated with highway construction in other areas. There has not been the disruption of social linkages, the constraining effect upon government activity, or the unfavorable design characteristics experienced elsewhere. Consequently, the respondents are not fully aware of the unfavorable aspects of highway construction. In some cases this is not entirely true, especially where open space considerations are involved. However, for the most part, the respondents tended to emphasize those spin-off effects from highway construction with which they have had past experience. As a result, economics, health and safety, and land use assume a dominant position in the region's responses. This, however, does not necessarily imply that these are the present concerns of the area.

A second cause of the observed response pattern may be associated with the strong regional desire for access and mobility as opposed to the purely nonuser effects of highway construction. This concern with the flow aspects of highways makes the respondents acutely aware of those community impacts that related to the safety on adjacent highways or the accessibility of emergency facilities. This may help to explain the emphasis in general placed upon the health and safety category. This characteristic, when combined with the one noted previously, makes it hard to determine whether the respondents were actually imply-

Table 5-1
Weighted Coefficient and Order of Rank for Each of the Thirty Potential Impacts (Group: South Central Connecticut Regional Planning Area)

Item Rank	Detrimental	Category	Weight	Beneficial	Category	Weight
1	Number of Jobs	3-A	.0503	Health Effects of Pollution	6-C	.0530
2	Health Effects of Pollution	6-C	.0499	Number of Jobs	3-A	.0517
3	Accessibility of Emergency Facilities	6-A	.0481	Accessibility of Emergency Facilities	6-A	.0488
4	Property Values	3-C	.0455	Level of Income	3-E	.0478
5	Safety on Adjacent Highway	6-E	.0442	Safety on Adjacent Highway	6-E	.0460
6	Level of Income	3-E	.0435	Property Values	3-C	.0445
7	Number of Business Firms	5-B	.0420	Amount of Open Space	5-C	.0425
8	Pattern of Land Development	5-A	.0411	Aesthetic Effects of Pollution	2-E	.04194
9	Amount of Open Space	5-C	.0384	Pattern of Land Development	5-A	.04193
10	Aesthetic Effects of Pollution	2-E	.0382	Number of Business Firms	5-B	.0397
11	Personal or Group Stress	6-D	.0378	Personal or Group Stress	6-D	.0392
12	Aesthetic Value of Right-of-Way	2-C	.0360	Community Cohesiveness	7-A	.0361
13	Blend Highway into Background	2-B	.0340	Number of Housing Units	5-D	.0352
14	Number of Welfare Recipients	3-B	.0339	Number of Welfare Recipients	3-B	.0341
15	Number of Housing Units	5-D	.0328	Aesthetic Value of Right-of-Way	2-C	.0341
16	Visual Quality of Highway	2-A	.0323	Blend Highway into Background	2-B	.0335
17	Community Cohesiveness	7-A	.0320	Community Oriented Contacts	7-E	.0313
18	Financial Capability of Government	4-C	.0310	Municipal Services	4-A	.0299
19	Municipal Services	4-A	.0295	Visual Quality of Highway	2-A	.0287
20	Neighborhood Stability	7-C	.0289	Financial Capability of Government	4-C	.0284
21	Community Security	4-D	.0283	Barrier Effect	7-D	.0268
22	Number of Historic Sites	5-E	.0277	Personal or Business Contacts	7-B	.0261

Table 5-1 (cont.)

Item Rank	Detrimental	Category	Weight	Beneficial	Category	Weight
23	Barrier Effect	7-D	.0271	Neighborhood Stability	7-C	.0260
24	Community Oriented Contacts	7-E	.0265	Community Security	4-D	.0240
25	National Defense	6-B	.0243	Public Participation in Government	4-B	.0210
26	Personal or Business Contacts	7-B	.0241	Number of Historic Sites	5-E	.0197
27	Public Participation in Government	4-B	.0224	Temporary Economic Effects	3-D	.0197
28	Temporary Economic Effects	3-D	.0219	National Defense	6-B	.0184
29	Satisfaction with Government	4-E	.0148	Temporary Aesthetic Effects	2-D	.0152
30	Temporary Aesthetic Effects	2-D	.0119	Satisfaction with Government	4-E	.0124
Category Rank	**Detrimental**	**Category**	**Weight**	**Beneficial**	**Category**	**Weight**
1	Health and Safety	8-E	.2045	Health and Safety	8-E	.2056
2	Economics	8-B	.1953	Economics	8-B	.1981
3	Land Use	8-D	.1823	Land Use	8-D	.1801
4	Aesthetics	8-A	.1525	Aesthetics	8-A	.1536
5	Social-Psychological	8-F	.1387	Social-Psychological	8-F	.1464
6	Political	8-C	.1263	Political	8-C	.1159
General Rank	**Detrimental**	**Category**	**Weight**	**Beneficial**	**Category**	**Weight**
1	Nonuser Effects	1-B	.5799	Nonuser Effects	1-B	.5128
2	User Effects	1-A	.4200	User Effects	1-A	.4871

ing that they were willing to trade off the social and governmental effects of highways in favor of economics or health and safety. Is it the desire for mobility or the lack of unfavorable past experience that dominates the response pattern?

A third possibility is that the questionnaire itself tended to present concepts which not only were of questionable relevance to the south central region, but which were also so general that with the limited experience of the area with these items, it was hard for the respondents to properly conceive of these items as being a threat in their particular town. In particular, the questionnaire may have appeared obscure in some respects resulting in a doubtful level of validity being attached to the replies. It is evident that some people in this region do not think of a highway in the terms described in the questionnaire. Although it is correct to point out the relative nature of highway impacts, presenting these particular items in the generalized questionnaire format may well have tended to distort the thinking the participants had done on these topics.

In retrospect, the pattern of response does seem to reflect the attitudes of the average resident of the area. However, as qualified by the three points noted above, the responses from a random sampling would also tend to reflect this lack of familiarity or concern with the items listed. In these terms, the sample does mirror the attitudes of their individual towns. It may also be that this pattern of response reflects a fourth consideration, namely, that the future of highway construction in Connecticut does not include many major plans for the south central region. As a result, the respondents were really operating in two levels of abstraction with a favorable past history and only a limited possibility of having any of these unfavorable effects being immediately visited upon their region.

Some variation in the pattern of response could be anticipated among the towns in the area. In particular, the responses obtained in the core city of New Haven would place far greater emphasis upon the social and individual impacts that a highway might have upon the community. Specifically, the elements of community cohesiveness and social linkages are of prime importance to urban residents, with housing and other aspects of land use following in second position. This would be consistent with the known characteristics of modern urban living. In general, by focusing upon the urban core, the highway planner would be obtaining a more accurate picture of what the true needs of the region are. There are two key problems which deserve particular attention. The first involves the nature of the core city itself which is undergoing significant changes in its social structure. These changes include alterations in the pattern of both residence and employment with the permanent out-migration of the middle class into the surrounding towns. This change of living patterns leaves the central city with social and financial difficulties which require external assistance in order to be properly corrected. The solution of this problem leads to the second issue which involves the further need for interaction between the towns in the region. Further cooperation between the urban-suburban areas are necessary if the central city is to solve the problems which are not totally of its own making. Consequently, highway planners in the area should take notice of these issues and if possible, develop avenues of mobility which will allow a greater level of cohesiveness between elements in the region.

Analysis of the Detrimental Results

The high level of concern associated with the economic effects of highway construction is evidenced by the premier rating assigned to the effects which the facilities might have upon the number of jobs. This, along with the sixth ranking associated with the effect on the level of income, has been partially explained in the opening paragraph. However, it is worth repeating that the past history of highway construction in the region has primarily resulted in economic impacts; consequently, these are the most real to the respondents. As a result, they would be capable of developing meaningful examples of them in their own town relative to the general statement provided in the questionnaire. Employment is not a severe problem in this region and the emphasis upon the harmful effects upon jobs cannot be associated with any of the hypotheses listed for other regions. In particular, past highway construction has increased rather than reduced the number of jobs. The composition of the commission is skewed in the direction of business concerns, consequently, since employment is dispersed throughout all of the towns in the area, the respondents might have been fearful of removing jobs from their town and transferring them to another town in the region. However, this has not been the effect of past highway construction, nor has there been a harmful effect upon the level of income. In particular, with respect to both of these factors, if anything, the opposite is true.

In the health and safety category, the answers obtained are more readily explainable by known characteristics of the region. The second ranking assigned to the health effects of pollution is consistent with the present national concern expressed regarding this issue and the fact that there are sources of air and water pollution in the area. These two points, when combined with the natural threat that the item itself poses for the individual, would seem to indicate this was a valid concern in the south central area. The access of emergency facilities ranked third and this too seems plausible. The pattern of both I-91 and I-95 tend to cut one part of the city off from the other with few ways provided by which emergency vehicles can move. There can be stretches of both which act as barriers to local traffic flow on urban streets and which set up the potential for crisis situations.

Continuing in the health and safety category, the fifth ranking assigned to the level of safety on adjacent highways represents an issue which has, in the past, been of primary concern to area residents. There are enough examples in the region of safety problems and sources of congestion to warrant this apparent high level of interest. In particular, the added traffic burdens placed upon local streets due to exit and entrance ramps has for a long time been a major issue. It has been pointed out that these urban streets were not constructed in a way which would allow for the safe and adequate handling of the added traffic flows.

In the land use category, the emphasis placed upon several of the items is partly explainable by past events. In particular, the ninth rank assigned to open space is related to the history of specific routes which were either planned or actually traversed open space territory. Existing open space comprises only approximately one-sixth of the present area of the region with most of this being concentrated in the Woodbridge, Bethany, Guilford, Madison, and North Branford areas. The urbanized regions of New Haven and Meriden have limited amounts of open space available and it is in these towns that conflicts with open space have predominated. This concern exists to the extent that private court cases have been brought by individuals regarding the use of park lands for highway construction. The current major source of complaint is the Hubbard Park area in Meriden, where conservationists claim that the construction of Route 66 may damage both the natural beauty and the wildlife sanctuary qualities of the area. Other conflicts have arisen concerning the proposed flow of traffic through the East Rock Park section of New Haven and the effect that the extension of the Oak Street connector will have on the West River and Memorial Park sections of the town.

The seventh rank assigned to the effect upon the number of business firms reinforces the concern with economic effects of construction. While the region as a whole has benefited significantly from past highway construction, the respondents, in being asked to reply with their specific towns in mind, may have been concerned about the movement of firms from one town to another within the region affecting both the employment base and the tax capabilities of the towns involved.

Continuing in the land use category, the low fifteenth rank assigned to housing again raises elements of skepticism. The region is experiencing a housing problem, although in several towns the concern is with the distribution of available houses rather than with the general level of availability. New Haven especially is experiencing a housing shortage in the low and middle income brackets. However, the general attitude of suburban residents concerning their willingness to disregard the problems of the central city may well have accounted for this particular item being ranked the way it was. Furthermore, the past history of adequate planning in this area has resulted in the avoidance of construction in high density residential neighborhoods. Consequently, this favorable experience may also have reduced the importance of the item in the minds of the respondents.

Switching to the aesthetic category, the concern expressed regarding the design aspects of blending the highway into the background and the aesthetic value of the right-of-way taken can partially be attributed to the concern already noted with the effect that highway construction has had on open space. In particular, past construction problems have lead to a conflict between the desire for mobility on the part of highway users and the effect that this added mobility will have on the quality of the environment in which residents must live. Moreover, certain high rise or elevated aspects in the design of the interstate highways in the region may well have generated concern for these factors in future construction.

The items in the political category receive comparatively little attention, as evidenced by the eighteenth and nineteenth rankings assigned to the financial capabilities of government and the effect upon the level of municipal services. This may be partially explained by the already noted low level of concern with nonuser items in general and the past favorable history of beneficial effects which have developed out of highway construction in the region. However, considering the severe problems being faced by many of the suburbs regarding the quality of their tax base and the high taxes which must be paid by property owners in the area, it seems odd that this item did not receive greater attention from the respondents.

Finally in the social-psychological category, the limited attention devoted to the particular items found here is characteristic of the region in that these factors are of little or no value to the areas surrounding the core city. Past highway construction has not led to the destruction of either the characteristics of neighborhoods or the development of strong social barriers which have segregated parts of the city in a socially destructive way. Although the highways do act as a barrier to traffic flows in certain places, the adequacy of preconstruction planning insured that the barriers which do exist did not significantly disrupt existing social linkages. Specifically, it would be hard to point out an instance where highway construction has substantially cut a neighborhood or adversely affected the existing pattern of community development.

Analysis of the Beneficial Results

The previous extensive examination of the response pattern coupled with the high degree of symmetry between the answers obtained in the beneficial and detrimental sides results in an offering of similar hypotheses to explain the rankings found on the beneficial side. In particular the primary concern with economic benefits as evidenced by the second ranking assigned to the effect upon the number of jobs and the fourth ranking assigned to the effect on the level of income, again mirrors the beneficial past history of economic spin-off effects from highway construction. Both I-91 and I-95 provided jobs and other developmental opportunities in the towns traversed with even the New Haven area experiencing a net addition in employment due to the development of the Long Wharf industrial park. As a result, respondents are highly concerned with furthering the economic development of the area which they see as a normal offshoot of highway construction. This is still true even today although it would appear that conditions have changed markedly in the New Haven area so that benefits other than those related directly to economics, should be the ones receiving the most attention from highway planners.

In the health and safety category, the concern with items which tend to affect the individual again appear prominently in the ranking. This is shown by the fact that the health effects of pollution, the access to emergency facilities, and safety on adjacent highways ranked first, third and fifth in the hierarchy. In retrospect, it's quite clear that there is a high degree of symmetry between the major items on both the beneficial and detrimental sides. This symmetry extends to the point that the first six items on the detrimental side again appear as the first six items on the beneficial side with minor variation in ordering.

These health and safety results can again be attributed, at least in part, to characteristics of the region. The desire for better traffic flows and safety on local streets is a justifiable one and rates appropriate merit from the respondents. Also, the need to improve the movement of emergency vehicles from one end of the city to the other in times of crisis is also justified owing to the tendency for the limited access highways in the area to serve as hindrances to localized cross traffic flows. Finally, the concern with the health effects of pollution appears as a reaction to the current state of national concern with this item, although there are cases where this too is a justifiable problem in the south central region. It would appear that in general the entire set of responses obtained to the items in this category would have some bases in either known facts or the characteristics of the region.

In the land use category the items of effect upon open space and the pattern of land development again received considerable attention being ranked seventh and ninth, although the importance attached to land development patterns in this area relative to that in other regions makes it appear slightly unimportant. This particular response may again be traced to the past history of adequate planning and the fact that except in a limited number of cases land development and highway construction have been mutually agreeable in their impact. The concern with open space is a result of the one-sixth allocation of existing land to open space and the desire on the part of planners to raise this to approximately one-third. Any highway construction which could assist in this open space development and conservation movement would be highly desirable.

Housing does not achieve as high a level of concern as it does in other urban regions, being placed thirteenth. However, as noted earlier, this may be due to the fact that the housing problem is not related to the availability of housing in terms of absolute numbers, but the fact that most of the housing is not readily accessible to low income groups. Consequently, the problem is not one of generating more houses as a whole, but more houses in specific income groups. Also, the suburban business orientation of the sample may have entered here by having them place a low priority on the construction of low income houses.

In the aesthetic category, the respondents showed approximately an average level of concern for all of the design characteristics with an emphasis placed only upon the aesthetic effects of pollution which ranked eighth. Here, quite rightly, the respondents were probably referring to the need for a general beautification of the south central area with specific reference to some sections in the town of New Haven. This would be especially true in areas traversed by I-91, which makes many of the least desirable sections of the region open to the full view of those who live and must drive through the region. This is a known problem and has generated quite a bit of attention in the past and will probably continue to do so in the future.

The final group worthy of note is the collection of items contained in the social-psychological category. Here, the respondents indicated general unconcern with these factors except for the cases of community cohesiveness and community oriented contacts which ranked twelfth and seventeenth in the ordering. Given the accuracy of the results already noted earlier, it may be hard to offer any tentative hypothesis regarding these particular responses. However, it is possible that the participants were aware of the need for a further unification of programs within the towns and throughout the region as a whole. As a result, they gave community cohesiveness, which refers to actions and goals, a high rating indicating there is some importance of this problem in the region. The importance attached to community oriented contacts may be a further manifestation of this need for a unification of purpose and a healthy desire on the part of members of the region to face some of the social problems which exist in the area and to work together to solve them.

5.3 Analytical Profile of the Midstate Regional Planning Area—A Semiurban Example

The Midstate Regional Planning Area is composed of the following eight towns: Cromwell, Durham, East Haddam, East Hampton, Haddam, Middlefield, Middletown, and Portland. The number of valid responses obtained in this region was thirteen. In general, the tabulated responses indicate that a high degree of regional concern is attached to the potential economic effects of highway construction. This interest is partially modified by environmental considerations, especially as they relate to current issues of a local nature within the planning region. These special concerns are presently directed towards the impact upon open space and the effect that a highway may have upon the level of pollution within the community. The respondents also showed a significant amount of concern for those items which dealt with the relationship between the individual and the government. On the other hand, little or no importance was attached to the items in the social-psychological and aesthetic categories with the latter grouping showing only a marginal improvement as the rating assigned to specific items moved from the detrimental to the beneficial side.

The responses appear to give a fairly accurate picture of the attitudes existing in the region relative to highway construction. The rankings are most applicable in terms of the high and low items. There is some feeling that the weights tend to overrepresent the majority view. As a result, certain impacts that cause social changes, which are necessary in the area, are neglected by the relative weights. Specifically, the sample, as in other regions, possesses a bias toward the suburban or semirural middle-class elements in the area. Consequently, many of the concerns evident within the core city of Middletown are afforded lesser attention. These urban problems which require special consideration include environmental health, social integration, the construction of utilities, and the solution of the housing problem.

Table 5-2
Weighted Coefficient and Order of Rank for Each of the Thirty Potential Impacts (Group: Midstate Regional Planning Area)

Item Rank	Detrimental	Category	Weight	Beneficial	Category	Weight
1	Health Effects of Pollution	6-C	.0485	Number of Jobs	3-A	.0526
2	Amount of Open Space	5-C	.0474	Level of Income	3-E	.0488
3	Level of Income	3-E	.0437	Health Effects of Pollution	6-C	.0480
4	Property Values	3-C	.0425	Accessibility of Emergency Facilities	6-A	.0474
5	Number of Jobs	3-A	.0423	Safety on Adjacent Highway	6-E	.0468
6	Pattern of Land Development	5-A	.0422	Property Values	3-C	.0447
7	Accessibility of Emergency Facilities	6-A	.0417	Number of Business Firms	5-B	.0426
8	Safety on Adjacent Highway	6-E	.0416	Personal or Group Stress	6-D	.0409
9	Number of Historic Sites	5-E	.0403	Pattern of Land Development	5-A	.0408
10	Number of Business Firms	5-B	.0381	Amount of Open Space	5-C	.0403
11	Personal or Group Stress	6-D	.0350	Financial Capability of Government	4-C	.0356
12	Aesthetic Effects of Pollution	2-E	.0342	Number of Housing Units	5-D	.0349
13	Community Security	4-D	.0341	Municipal Services	4-A	.0335
14	Financial Capability of Government	4-C	.0337	Community Security	4-D	.0319
15	Number of Housing Units	5-D	.0332	Number of Welfare Recipients	3-B	.0314
16	Municipal Services	4-A	.0328	Number of Historic Sites	5-E	.0313
17	Blend Highway into Background	2-B	.0327	Aesthetic Aspects of Pollution	2-E	.0292
18	Aesthetic Value of Right-of-Way	2-C	.0321	Public Participation in Government	4-B	.0289
19	Number of Welfare Recipients	3-B	.0316	Community Cohesiveness	7-A	.02838
20	Community Cohesiveness	7-A	.0311	Neighborhood Stability	7-C	.02832
21	Visual Quality of Highway	2-A	.0303	Community Oriented Contacts	7-E	.0275
22	Community Oriented Contacts	7-E	.0298	Aesthetic Value of Right-of-Way	2-C	.0270

Table 5-2 (cont.)

Item Rank	Detrimental	Category	Weight	Beneficial	Category	Weight
23	Neighborhood Stability	7-C	.0291	Visual Quality of Highway	2-A	.02505
24	Public Participation in Government	4-B	.0284	Barrier Effect	7-D	.02504
25	Satisfaction with Government	4-E	.0258	Blend Highway into Background	2-B	.0248
26	Barrier Effect	7-D	.0248	Temporary Economic Effects	3-D	.0243
27	National Defense	6-B	.0241	Personal or Business Contacts	7-B	.0227
28	Personal or Business Contacts	7-B	.0230	National Defense	6-B	.0220
29	Temporary Economic Effects	3-D	.0131	Satisfaction with Government	4-E	.0210
30	Temporary Aesthetic Effects	2-D	.0113	Temporary Aesthetic Effects	2-D	.0128

Category Rank	Detrimental	Category	Weight	Beneficial	Category	Weight
1	Land Use	8-D	.2014	Health and Safety	8-E	.2054
2	Health and Safety	8-E	.1910	Economics	8-B	.2020
3	Economics	8-B	.1735	Land Use	8-D	.1902
4	Political	8-C	.1549	Political	8-C	.1511
5	Aesthetic	8-A	.1409	Social-Psychological	8-F	.1320
6	Social-Psychological	8-F	.1380	Aesthetic	8-A	.1190

General Rank	Detrimental	Category	Weight	Beneficial	Category	Weight
1	Nonuser Effects	1-B	.4816	User Effects	1-A	.5146
2	User Effects	1-A	.5183	Nonuser Effects	1-B	.4853

In general, the most significant source of variation in the answers supplied appeared as the administration of the questionnaire moved from the urban center out toward the more rural towns in the planning area. In the southern half of the region, the concern of the towns is directed more toward the land use aspects of highway construction with some emphasis upon economics in those areas which possess a significant amount of daily out-migration for employment purposes. In these areas, the social items receive the least amount of attention with progressively more devoted to them as the proximity to Middletown increases.

Analysis of the Detrimental Results

On the detrimental side, the most important item was the health aspects of pollution. This appears to be a justifiable concern which has attracted considerable attention in the past in this region. First, the area, on a countywide basis, is presently engaged in a federal program of comprehensive health planning which is aimed at developing a coordinated health plan for the region as a whole. This program stimulates individual interest by the use of the citizen advisory board technique, which directly involves local residents making them fully aware of the health problems facing the separate towns.

Secondly, several towns have been operating under state orders to develop or expand their sewage treatment facilities to avoid problems associated with the contamination of the Connecticut River. As a result, the towns involved have recently engaged in the construction of sewage treatment plants, with both the existence and the cost of this operation being well-known to the representatives from the area.

Finally, these two facts when combined with the current amount of attention directed to the problem of environmental pollution in the nation as a whole, seems to provide a justification for the observed level of concern attached to this particular item. It is interesting to note that this concern with environmental quality does not appear to be an emotional reaction to the word "pollution" as the entry referring to the aesthetic effects of pollution ranks only twelfth in the hierarchy. Obviously, the respondents were able and willing to make a conscious separation in the values attached to the two items.

The second rank attached to the item of open space may have been a reaction to the proposed corridor location or Route 66 in the Middletown area. Specifically, the construction of this highway will mean the elimination of a small vest pocket recreational area known as Ravine Park. This park is of considerable importance to the homeowners located in the surrounding area. As a result, the preservation of the facility has generated a high level of topical interest in the community.

The importance of this issue raises two critical questions. The first regards the dispersion of preserved open space which does amount to almost 20 percent of the land area in the midstate region. The second involves the trade-offs which must be established between the values of the city as a whole as opposed to the concerns of a localized neighborhood group. Concerning the first item, the Middletown area possesses little of the available open space that has already been preserved throughout the region. Most of this open space is concentrated in the outlying communities, with the towns of Haddam and Hampton containing sizeable amounts. Also, the town of Durham has recently engaged in an active open space program, which culminated in the public purchase of over 1,000 acres that were immediately earmarked for conservation purposes.

In comparison to this apparent abundance of suburban open space facilities, Middletown, with limited park and recreational facilities presently available, is highly concerned with the destruction of a facility which provides adjacent open space to a residential area. The issue is that, although the park itself can be replaced by recreational construction in another part of the town, the proximity of the park to the residents of the neighborhood would no longer be present. Responsible elements in the town have recognized the necessity of taking the park, and actually advocated its removal since the alternative routes would necessitate the destruction of homes and the displacement of numerous residents. Since the construction of Route 66 in the area is a desirable item in itself, the impact trade-offs of homes versus open space is clearly the issue at point. The city or regional point of view favors the maintenance of housing and is opposed to the set of neighborhood values which are intent upon maintaining the existence of the park within easy traveling distance. It is the relative importance of these two items and points of view which clearly demands the center of attention in this dispute.

The next set of concerns are economically oriented as evidenced by the ratings assigned to the level of income, the effect upon property values and upon the number of jobs which ranked third, fourth and fifth in the hierarchy. These reponses are consistent with the known character of the area, since the net outmigration approaches 20 percent of the labor force. Consequently, any highway construction which would further aggrevate this situation by destroying jobs already existing in the area should be avoided. This concern with economics is also consistent with the low tenth ranking assigned to the number of busines firms. This apparent paradox exists since many of the suburban communities do not possess a manufacturing base. As a result, while concerned over the destruction of employment opportunities in general, the respondents did not see a heavy threat to the existence of industry within their own towns. The only areas where highway construction could significantly affect the industrial base would be in the towns of Portland and Middletown, where most of the industry is concentrated.

Continuing in the land use category, the respondents in the midstate region were among the few who attached a high level of importance to the number of historic sites. This item ranked twelfth in the ordering and may have been caused by the extensive colonial heritage which exists in many of the suburban towns including Haddam, Durham, and East Hampton. The residents of these towns are aware of the existence of these sites and they regard them as a local asset, which improves the quality of the region by attracting desirable new residents into the area. Such sites are considered to be a cultural heritage and as indicated, they may be more important to the nature of the individual town than either the effect upon business or the impact upon housing in general. These sites are regarded as being the focal point of the community and their value is quite high in providing the towns with their respective character.

Housing receives a relatively low rank, being placed fifteenth, since most of the housing problem is concentrated in the greater Middletown area. Moreover, the respondents sampled were in general a homeowning suburban class of participants. In comparison to other regions, the low income housing problem is of average importance. However, the worst housing shortages actually exist among the homes available for middle class residents. Within the Midstate Region the chances for solving the low income housing problem are greater than those for alleviating the middle class housing problem, since the former will attract considerable Federal support in terms of finance.

The aesthetic category attracted little attention from the respondents with the most important item being the aesthetic effects of pollution which ranked 12th. Those items which referred to design aspects of highway construction were judged to be of conditional importance, being placed seventeenth, eighteenth and twenty-first. This may be accounted for, in part, by the lack of a past history with aesthetically undesirable design problems. The most recently completed project in the area was Route 9, which was not harmful aesthetically to the midstate region. If anything, the design enhances the beauty of the region by opening it for public view as the motorists travel through the area. Any design issues which do exist are related to highways that have been on the ground for several years. Since these design problems were not repeated with the construction of new highways, the respondents are less concerned with the aesthetic effects of design than they might be in other regions.

In the political category, the above average attention directed toward the financial capabilities of government and the effect upon the level of municipal services can partially be attributed to the suburban nature of the sample. While these items ranked fourteenth and sixteenth in the ordering, the principle concern with them exists in the outlying areas. Although several towns in the region possess tax rates which are among the lowest in the state, there is a sufficiently heavy tax burden in most of the towns to warrant the attention directed toward this item. With respect to services, many of the smaller towns actually provide little or no municipal services; consequently, the respondents might have been concerned with avoiding any disruption of those that are provided and are considered to be the minimum essential amount for adequate living. The item of community security also attracted considerable attention being placed thirteenth. This may be in part a reaction to the civil disorders which have appeared periodically in the core city. Moreover, the respondents may have been fearful that the development of greater access will result in the inflow of additional persons into the community which will tend to disrupt the existing structure of the individual town.

Finally, the social-psychological category, received very little emphasis. For the most part this is a genuine indication of the lack of concern with personal communications in the area. Specifically, respondents are not overly concerned with the types of social linkages which are developed in the more rural communities. This set of responses may also be a manifestation of their past experience with highway construction. Neither Route 66 nor Route 9 have any significant effect upon the social organization of the communities traversed. No segments of towns were separated from each other nor were important communication linkages disrupted due to the location of these highways.

Analysis of the Beneficial Results

On the beneficial side, the respondents were even more intensely concerned with the economic spin-off effects of highway construction. This is evidenced by the first and second rankings assigned to the effect upon the number of jobs and the level of income. Moreover, this tendency is reinforced by the set of rankings assigned to the effect upon the number of business firms. This desire for a potential growth stimulus in the region can be accounted for by two facts. First, there is a strong desire for an increase in the employment base of the area. It is important to the respondents to provide more jobs within the region in order to reduce the flow of out-migration for employment purposes to other regions. Secondly, it is possible to relate this concern for an increased value attached to the financial capabilities of government which ranked eleventh in the ordering. It is clear that in some towns the people are making a connection between attracting new industry, which will help support the tax base, and the increased number of employment opportunities that this industry will bring into the area. As noted earlier, this concern will vary somewhat throughout the region, with some towns feeling a significant tax burden upon the individual, while others are relatively well off taxwise compared to other communities in both the region and throughout the state.

In the land use category, the items received a varying amount of attention, with some emphasis placed upon the potential effect that a highway may have on the number of historic sites. Placing this item sixteenth in the hierarchy can again be attributed to the historical nature of many of the surrounding communities. This point, when analyzed relative to the suburban nature of the sample, may tend to distort the importance of this impact throughout the region as a whole. Specifically, it should not be expected that Middletown would emphasize the importance of historic sites to the degree as in Haddam, Durham, or East Hampton. The item of open space is ranked tenth and does not receive as much relative attention in this as opposed to other regions. This is consistent with the fact that many of the towns already devote a considerable amount of their available land to open space preservation. Consequently, highways which add to this available amount of conservation land are not as important with respect to this item as they would be in other regions. Finally, in the land use category, the ninth ranking associated with the pattern of land development is partially attributed to the favorable past history of highway construction in the region. The construction of Route I-91 lead to the development of industrial parks which made proper use of available land. This point when combined with the ready availability of land for further development purposes, tends to make the residents of the region less concerned about benefiting the pattern of land development than they might be in other regions.

In the aesthetic category, the low level of concern attached to the design items which were ranked twenty-second, twenty-third, and twenty-fifth is again the result of the lack of an unfavorable past history with design in the region. Furthermore, it would appear that even where the aesthetic effects of pollution are concerned, which ranked seventeenth, the respondents were indicating that there are other potential beneficial effects of far greater importance, which can significantly improve the region.

Finally, in the social-psychological category, the respondents did not consider these items of any great importance to them. The element of community cohesiveness, which ranked nineteenth, attracted the most attention. However, even it was ranked low relative to the position assigned to this impact in other regions. This indicates that developing social or community linkages beyond those which already exist in the community should not be a variable which is given significant weight in the corridor selection process. This pattern of response may be due to the suburban nature of the sample with the answers tending to vary as the location of the route moves closer to the core city of Middletown, where neighborhood or community items may appear to be more vital. An example of this may well exist in the interest generated by taking of Ravine Park. The highly localized community felt that the social as well as the recreational basis of their neighborhood was to be affected by the removal of this park.

5.4 Analytical Profile of the Litchfield Hills Regional Planning Area—A Rural Example

The Litchfield Hills Regional Planning Agency is composed of the following nine towns: Barkhamsted, Goshen, Harwinton, Litchfield, Morris, New Hartford, Norfolk, Torrington, and Winchester. The number of valid responses obtained from this area totaled fifteen. The pattern of response indicates that the participants were concerned primarily with the economic and land use effects of highway construction, modified by particular concerns in the health and safety category. There was some element of symmetry among the rankings assigned to the top items considered as most important on both the beneficial and detrimental sides.

Beyond the items of dominant concern there is evidence of a considerable variation in the position assigned to the items located in the lower two-thirds of the hierarchy. This variation appears most acute with respect to the items found in the social-psychological category which seem more important to the respondents as they consider them from the detrimental as opposed to the beneficial sides. This could be a tentative indication of a satisfactory existing social structure which highway construction should avoid disrupting to any great extent. Minor fluctuations are also observed in the aesthetic and political categories, with the majority of these items appearing more important on the beneficial side. For the latter group of items, this would show an area wide need for the further development of a governmental structure which would be better able to

Table 5-3
Weighted Coefficient and Order of Rank for Each of the Thirty Potential Impacts (Group: Litchfield Hills Regional Planning Area)

Item Rank	Detrimental	Category	Weight	Beneficial	Category	Weight
1	Health Effects of Pollution	6-C	.0490	Number of Jobs	3-A	.0452
2	Property Values	3-C	.0468	Health Effects of Pollution	6-C	.0451
3	Number of Jobs	3-A	.0459	Safety on Adjacent Highway	6-E	.0436
4	Accessibility of Emergency Facilities	6-A	.0448	Number of Business Firms	5-B	.0435
5	Pattern of Land Development	5-A	.0436	Level of Income	3-E	.0434
6	Safety on Adjacent Highway	6-E	.0431	Pattern of Land Development	5-A	.0424
7	Number of Business Firms	5-B	.04026	Property Values	3-C	.0423
8	Level of Income	3-E	.04020	Accessibility of Emergency Facilities	6-A	.0418
9	Number of Housing Units	5-D	.0381	Amount of Open Space	5-C	.0385
10	Aesthetic Effects of Pollution	2-E	.0371	Personal or Group Stress	6-D	.0371
11	Personal or Group Stress	6-D	.0371	Number of Housing Units	5-D	.0368
12	Amount of Open Space	5-C	.0369	Temporary Aesthetic Effects	2-E	.0366
13	Number of Welfare Recipients	3-B	.0354	Number of Welfare Recipients	3-B	.0357
14	Community Oriented Contacts	7-E	.0345	Aesthetic Value of Right-of-Way	2-C	.0339
15	Neighborhood Stability	7-C	.0331	Municipal Services	4-A	.0301
16	Aesthetic Value of Right-of-Way	2-C	.0330	Community Cohesiveness	7-A	.0300
17	National Defense	6-B	.0314	National Defense	6-B	.02949
18	Community Cohesiveness	7-A	.0311	Visual Quality of Highway	2-A	.02947
19	Barrier Effect	7-D	.0310	Temporary Economic Effects	3-D	.02936
20	Blend Highway into Background	2-B	.0283	Blend Highway into Background	2-B	.02933
21	Personal or Business Contacts	7-B	.0280	Financial Capability of Government	4-C	.028852
22	Visual Quality of Highway	2-A	.0264	Neighborhood Stability	7-C	.028851

Table 5-3 (cont.)

Item Rank	Detrimental	Category	Weight	Beneficial	Category	Weight
23	Number of Historic Sites	5-E	.0258	Community Security	4-D	.0283
24	Temporary Economic Effects	3-D	.0254	Community Oriented Contacts	7-E	.0268
25	Financial Capability of Government	4-C	.0246	Barrier Effect	7-D	.0266
26	Municipal Services	4-A	.0242	Number of Historic Sites	5-E	.0263
27	Community Security	4-D	.0239	Personal or Business Contacts	7-B	.0240
28	Temporary Aesthetic Effects	2-D	.0214	Satisfaction with Government	4-E	.0222
29	Public Participation in Government	4-B	.0211	Public Participation in Government	4-B	.0220
30	Satisfaction with Government	4-E	.0170	Temporary Aesthetic Effects	2-D	.0213

Category Rank	Detrimental	Category	Weight	Beneficial	Category	Weight
1	Health and Safety	8-E	.2056	Health and Safety	8-E	.1972
2	Economics	8-B	.1939	Economics	8-B	.1961
3	Land Use	8-D	.1847	Land Use	8-D	.1877
4	Social-Psychological	8-F	.1579	Aesthetics	8-A	.1506
5	Aesthetics	8-A	.1465	Social-Psychological	8-F	.1972
6	Political	8-C	.1111	Political	8-C	.1316

General Rank	Detrimental	Category	Weight	Beneficial	Category	Weight
1	Nonuser Effects	1-B	.5161	User Effects	1-A	.5119
2	User Effects	1-A	.4838	Nonuser Effects	1-B	.4880

serve the needs of the local residents. On the other hand, the aesthetic concerns may show that the respondents are concerned with the proper use of their existing environment.

The sample of agency members, as supplemented by concerned members from several of the towns, appears to be fairly representative of the region as a whole. This would seem plausible since the Litchfield Hills area is among the smallest in population and consequently those people who feel that they are speaking as the voice of the larger community usually have a good indication of the attitudes of the overall population. In further support of the representative nature of the participating group, it appears that many of the rankings associated with particular items are true manifestations of characteristics of the region. All of the towns are small and fairly homogeneous in the make-up of their populations; therefore the points of interest as indicated by the questionnaire are held in common by all. Little variation could be expected as the questionnaire moved from town-to-town within the region. In most of the towns the pressing issues would include drainage and flood control, the problems associated with economic development, including the need for daily commuting to other regions in order to find employment, and problems regarding both the quality and quantity of available housing. In summary, the quality of the answers supplied in the region appears to be high in reflecting the community's attitudes toward future highway construction.

Analysis of the Detrimental Results

On the detrimental side, the emphasis placed upon the health aspects of pollution, which was rated first, appears to be a genuine concern in the area. This concern is a combination of both historical experience with the problems of pollution and the general familiarity with the current publicity given to the topic of environmental quality. The Naugatuck River, which runs through the area, has in the past exhibited heavy amounts of pollution which have tended to alert the area's residents to the need for water quality. Furthermore, the nature of the soil in the region generates significant problems with drainage facilities. Consequently, in addition to the interest attached to the recognizable level of air pollution which already exists in the area, a strong concern with water quality is also evident. This concern extends to the point that the Litchfield Hills region is one of only two areas in the state that possesses a regional health district. With the appointment of a full-time health director, there is an obvious recognition that the health of the area and the quality of the environment must be protected. Continuing in the health and safety category, the concern with safety on adjacent highways can be attributed to the topography of the area, which, with its rolling hills, generates severe traffic flow problems. While the major arteries are well designed, there is a tendency to divert traffic onto urban streets, which, at times, are not capable of handling the added flow.

Economic factors are of major importance, with property values and the effect on the number of jobs being ranked second and third. The concern with jobs can be traced to the fluctuating level of unemployment which is related to the highly seasonal nature of the employment base. A significant number of residents find part or all of their employment in the area's summer recreational facilities which comprise over six square miles of land and hire workers for less than six months of the year. On the other hand, many of those who are employed full-time receive a relatively low wage from the region's industry. These limited job opportunities necessitate a significant emigration of people into other areas in order to find daily employment. This concern with the economic condition of the area is further demonstrated by the high ranking assigned to the harmful effect that a highway may have on the number of business firms. Although the existing industrial base of the area is acceptable even if low paying, the seventh ranking assigned to this item demonstrates the natural concern with any displacement of industry which could leave a large number of available workers unemployed, further aggravating the job situation.

Continuing in the land use category, as might be expected, the respondents placed housing high on the list, ranking it ninth in the priority of items to be avoided by highway construction. This is a reflection of the deteriorated nature of housing in the area by census standards. Not only are these homes in poor condition, but they also possess a high degree of visibility further impressing upon the respondents the need to avoid added damage to the housing situation. In connection with this, it is interesting to note that while most of the area is rural in nature, the location of housing is quite dense with certain portions of Torrington, Winsted, and New Hartford containing the majority of available homes. Finally, the item of open space ranked twelfth placing it relatively low on the scale. This particular response is caused by the recreational camps which take up much of the land in the area with significant portions also being devoted to watershed holdings. These existing allocations of land to open space purposes make the respondents far less concerned with having a highway adversely affect the amount available in the area.

In the aesthetic category, the items found here are not given exceptional weight. This may be the result of two factors. First, due to the need for economic development, the highway is thought of as a force bringing further growth to the area, which is considered more important than maintaining purely aesthetic qualities. As an example, it should be noted that Route 8 was actually fought for by the residents as a means of opening up the region for economic development. Secondly, the area really does not have a history of unfavorable design or aesthetic problems with highways. Route 8 is again a good example; it blended well into the background and opened up the area to the visual benefit of persons traveling the road.

Regarding the items contained in the political category, the importance attached to them is as low in this region as it is in any other area. This is especially true of the financial capabilities of government and municipal services which ranked twenty-fifth and twenty-sixth on the detrimental side. This ordering can be traced to the already low level of municipal services provided by the local towns. The only major service outside of education is road maintenance. Consequently, except for minimal concern there is no real harm that can be done to either the quantity or quality of services provided in the area as the result of highway construction. This lack of concern regarding municipal services is translated to the effect upon taxes. Without an extensive level of services, there is no real need to protect the tax base or the financial capability of the government.

This is also consistent with the evidence that many of the services that are usually provided by the towns, such as welfare or community recreational items, are supplied in the Litchfield region on a private rather than a public basis. Consequently, there is not the heavy need for an inflow of funds to maintain governmental functions.

Finally, the items contained in the social-psychological category do receive greater attention in this as opposed to other regions. This may be due to the high degree of private social organization in the area. There appears to be a natural sorting out of residents into ethnic or economic groups with the majority of residents wanting to maintain this division along neighborhood lines. As a result, any highway which would disrupt the neighborhood stability of the area or the extent of community linkages would be highly undesirable. This neighborhood orientation extends partly to the town level as expressed by the concern given to community cohesiveness which ranked eighteenth. However, people in the area in general do not rely upon local government as much as they do in other regions.

Analysis of the Beneficial Results

On the beneficial side, it is evident that the respondents were primarily concerned with economic considerations. This is shown by the first and fifth rankings assigned to the effect on the number of jobs and the effect on the level of income. As indicated earlier, this interest can be attributed to a combination of three factors. The first is the highly seasonal nature of employment in the region which at times provides upwards of 7 percent unemployment during the winter months. The second is the heavy out-migration in the form of daily commutation trips to other regions to find employment. Finally, the third is the general low level of wages received by workers who are now employed in the region. As a result, of these considerations, the residents of the region are particularly concerned about attracting new industry with added jobs which will tend to raise the wage standard of workers. Further recognition of this concern for the employment base is indicated by the ranking assigned to the effect upon welfare recipients on both the beneficial and detrimental sides. In particular, the continued fear of possible unemployment and the low level of existing wage payments tends to make the participants more sympathetic with the standard of living of those who are unable to find adequate employment.

In the land use category, this concern with economics is further evidenced by the fourth position assigned to the beneficial effect that a highway may have on the number of business firms. However, in modifying this with the sixth position assigned to the pattern of land development, there is some indication that the area's residents are not willing to sacrifice totally, the quality of their environment. This concern with environmental quality is also reflected in the importance attached to the health effects of pollution, which was placed second. The safety on adjacent highways is another health and safety item which attracted considerable concern being placed third. This rank is due to the poor safety condition of many secondary roads in the area caused by the hilly topography of the region. Consequently, people fear aggravating this problem by placing added traffic burdens on these roads.

Finally, the items found in both the social-psychological and aesthetic categories are not given heavy consideration by the respondents. The only exception to this in the social-psychological category is the sixteenth rating assigned to community cohesiveness which may indicate a desire on the part of the people to have a greater town organization and closer connection to the local governing body. In the aesthetic category, the aesthetic value of the right-of-way is considered important by the participants and may be partially directed toward the removal of unsightly buildings that already exist in the region.

Presentation and Analysis of the Impact Factor Weights

6.1 An Overview of the Results

The following analysis, accompanied by the numerical tables, provides a detailed description of the variation that exists in the regional weights obtained during the survey phase of the study. It points out both the extent and the quality of the variation by centering upon a review of the fluctuation in the ordinal ranking of the items. These changes in rank mirror the range of deviation associated with the cardinal weights presented in sample form in Chapter 5 containing the analytical profiles. This ordinal analysis facilitates the comparison of the answers supplied by each of the participating groups, and allows the reader to grasp the essential differences that exist between clearly defined regional groups concerning their attitudes toward the potential impacts of highway construction.

In order to make the analysis as simple and uncluttered as possible, there are only two sets of tables presented in this section. A careful review, however, of the information contained within them should reveal the essence of the existing concerns held in each of the regions. The first set shows the rank order of appearance for each of the thirty items as they were evaluated by the fifteen participating groups. This presentation considers the impacts from both the detrimental and beneficial points of view. The second set of tables analyzes the observed variation in rank between the thirteen planning regions. This analysis concentrates on three pieces of information regarding each of the thirty items: (1) What was the extent of variation in the rank? (2) Over what range did the variation appear? (3) How important is a given item in the value structure of the participants? The results of the profile should provide the highway planner with a sound basis for establishing predictions regarding the importance of these items in future construction plans.

Presentation of the Ordinal Ranking

Table 6-1 shows the order of rank assigned to the hypothetical detrimental impact of each item classified according to the participating group. These groups are noted by the column headings, with the first thirteen entries indicating the responses obtained from the regional planning representatives. The first two columns to the right of the double line indicate the responses of the two supplemental groups, the highway planners and the regional planning directors. The

Table 6-1
Ordinal Ranking of the Detrimental Impact Items Categorized by Participating Group

Detrimental Items / Participating Groups	Capitol RPA	Central Connecticut RPA	Central Naugatuck Valley RPA	Connecticut River Estuary RPA	Greater Bridgeport RPA	Litchfield Hills RPA	Midstate RPA	Northeast RPA	South Central RPA	Southeastern RPA	Southwestern RPA	Valley RPA	Windham RPA	State Highway Officials	Regional Planners	State
Aesthetics																
1. Visual Quality of Highway	24	12	13	17	13	22	21	13	16	18	13	21	19	16	11	15
2. Blend Highway into Background	15	10	10	16	19	20	17	11	13	15	9	16	21	9	5	12
3. Aesthetic Value of the Right-of-Way	16	11	14	10	18	16	18	10	12	13	11	18	18	12	20	13
4. Temporary Aesthetic Effects	30	30	30	30	30	28	30	30	30	30	30	29	30	30	30	30
5. Aesthetic Effects of Pollution	14	8	17	3	7	10	12	14	10	7	3	8	5	15	7	9
Economics																
1. Number of Jobs	5	2	6	8	10	3	5	5	1	1	8	4	3	3	3	4
2. Number of Welfare Recipients	20	21	18	24	24	13	19	7	14	8	26	11	12	8	15	18
3. Property Values	1	3	3	6	5	2	4	4	4	2	5	3	2	2	9	2
4. Temporary Economic Effects	21	17	27	27	26	24	29	23	28	28	24	20	28	29	28	26
5. Level of Income	6	1	4	9	9	8	3	16	6	3	12	5	4	1	14	6
Political																
1. Municipal Services	10	18	11	19	14	26	16	19	19	22	25	14	17	20	22	17
2. Public Participation in Government	25	28	15	23	16	29	24	28	27	27	29	25	27	25	23	27

3. Financial Capability of Government	11	15	9	21	23	25	14	18	18	23	23	12	16	21	16	19
4. Community Security	18	26	12	26	21	27	13	22	21	26	22	9	23	24	27	22
5. Satisfaction with Government	29	29	29	29	29	30	25	29	29	29	27	30	29	28	24	29
Land Use																
1. Pattern of Land Development	7	13	7	4	11	5	6	2	8	12	10	17	8	4	19	7
2. Number of Business Firms	12	6	19	12	22	7	10	1	7	9	18	7	9	7	25	10
3. Amount of Open Space	8	16	8	2	3	12	2	12	9	10	2	13	15	6	17	8
4. Number of Housing Units	9	9	20	11	8	9	15	8	15	16	7	10	20	11	12	11
5. Number of Historic Sites	23	22	21	13	20	23	9	20	22	14	17	28	22	18	26	20
Health and Safety																
1. Access to Emergency Facilities	3	7	1	5	4	4	7	3	3	5	4	2	7	10	8	3
2. Effect on National Defense	27	20	26	28	25	17	27	17	25	17	28	26	24	26	29	28
3. Health Effects of Pollution	2	4	2	1	1	1	1	6	2	4	1	1	1	13	1	1
4. Personal or Group Stress	13	14	16	7	12	11	11	15	11	11	20	15	10	17	2	14
5. Safety on Adjacent Highways	4	5	5	15	2	6	8	9	5	6	6	6	6	5	6	5
Social-Psychological																
1. Community Cohesiveness	17	19	22	20	6	18	20	21	17	21	15	22	13	23	10	16
2. Personal or Business Contacts	26	25	28	25	27	21	28	25	26	19	19	24	26	27	21	25
3. Neighborhood Stability	28	27	24	18	15	15	23	24	20	24	16	27	14	22	4	23
4. Barrier Effect	22	24	25	22	17	19	26	26	23	25	21	23	25	19	13	24
5. Community Oriented Contacts	19	23	23	14	28	14	22	27	24	20	14	19	11	14	18	21

Table 6-2
Ordinal Ranking of the Beneficial Impact Items Categorized by Participating Group

Beneficial Items / Participating Groups	Capitol RPA	Central Connecticut RPA	Central Naugatuck RPA	Connecticut River Estuary RPA	Greater Bridgeport RPA	Litchfield Hills RPA	Midstate RPA	Northeast RPA	South Central RPA	Southeastern RPA	Southwestern RPA	Valley RPA	Windham RPA	State Highway Officials	Regional Planners	State
Aesthetics																
1. Visual Quality of Highway	22	16	19	27	25	18	23	12	19	17	18	22	19	20	19	19
2. Blend Highway into Background	18	12	12	17	21	20	25	15	16	14	15	21	15	13	13	16
3. Aesthetic Value of the Right-of-Way	16	9	13	15	19	14	22	9	15	4	14	10	14	15	20	13
4. Temporary Aesthetic Effects	30	30	30	30	30	30	30	28	29	28	30	28	30	30	28	30
5. Aesthetic Effects of Pollution	11	10	8	11	17	12	17	10	8	5	12	7	5	17	14	10
Economics																
1. Number of Jobs	3	1	2	10	1	1	1	2	2	2	6	1	1	3	1	2
2. Number of Welfare Recipients	12	17	11	16	10	13	15	8	14	8	26	5	13	6	10	14
3. Property Values	8	5	5	5	7	7	6	4	6	3	8	8	3	5	9	7
4. Temporary Economic Effects	27	19	28	29	29	19	26	21	27	23	27	24	25	28	30	28
5. Level of Income	1	4	3	9	2	5	2	7	4	7	9	2	2	1	6	3
Political																
1. Municipal Services	14	18	17	4	18	15	13	23	18	26	19	13	17	18	24	17
2. Public Participation in Government	24	21	22	14	24	29	18	26	25	29	29	23	21	27	25	25
3. Financial Capability of Government	17	15	18	7	14	21	11	19	20	24	23	12	12	12	18	18

4. Community Security	21	25	21	8	16	23	14	25	24	27	25	15	24	23	23	23
5. Satisfaction with Government	29	29	29	23	27	28	29	30	30	30	28	30	28	22	26	29
Land Use																
1. Pattern of Land Development	4	8	6	1	5	6	9	3	9	10	3	18	6	2	11	5
2. Number of Business Firms	9	6	15	12	13	4	7	1	10	13	11	9	8	8	17	9
3. Amount of Open Space	7	14	9	6	6	9	10	13	7	9	2	16	11	4	15	8
4. Number of Housing Units	10	11	16	22	8	11	12	16	13	20	7	14	23	16	4	11
5. Number of Historic Sites	28	24	26	20	28	26	16	29	26	19	13	29	27	29	27	26
Health and Safety																
1. Access to Emergency Facilities	6	7	4	3	3	8	4	5	3	11	4	4	7	10	7	4
2. Effect on National Defense	26	23	27	28	23	17	28	18	28	21	24	26	22	26	29	27
3. Health Effects of Pollution	2	3	1	2	4	2	3	6	1	1	1	3	4	9	2	1
4. Personal or Group Stress	13	13	14	19	9	10	8	14	11	12	10	11	10	14	3	12
5. Safety on Adjacent Highways	5	2	7	13	15	3	5	11	5	6	5	6	9	7	5	6
Social-Psychological																
1. Community Cohesiveness	15	20	10	18	11	16	19	17	12	15	20	20	16	19	8	15
2. Personal or Business Contacts	25	26	25	25	22	27	27	24	22	25	21	27	26	25	22	24
3. Neighborhood Stability	19	28	23	24	12	22	20	20	23	22	16	25	18	11	12	21
4. Barrier Effect	23	22	24	26	20	25	24	22	21	16	22	19	29	21	21	22
5. Community Oriented Contacts	20	27	20	21	26	24	21	27	17	18	17	17	20	24	16	20

latter set of answers was developed on a statewide basis through the use of weights derived by applying regional population percentages. The aggregating process utilized projected 1970 state and regional population estimates obtained from the Connecticut Interregional Planning Program and issued in December 1968. The final column shows the rank order of appearance assigned by the aggregate of regional respondents with the summation process again accomplished through the application of population weights. The row entries refer to the thirty items arranged in groups with their order of appearance the same as in the questionnaire. The numerical rankings themselves are derived from the weights tabulated in each of the participating groups. The items are ranked from top to bottom with the most important item receiving a ranking of one and the least important item, as indicated by its relative weight, receiving a rank of thirty. Table 6-2 repeats the format for both the participating groups and the potential impact factors, except that this rank ordering refers to the potential impacts as beneficial effects of highway construction.

The observations contained in the first set of tables are arranged and analyzed in table 6-3; they describe the extent of item place variation among the thirteen planning regions. The two supplementary groups are omitted from this calculation, since the participants were drawn from a separate universe and their answers were given from a slightly different point of view. Consequently, in many cases, the weights obtained from these groups form an extreme position outside the range of answers supplied by the planning regions. As a result, their inclusion might tend to distort the analysis and turn the presentation away from its original purpose of assessing the value of public participation in corridor route locations. The presentation of table 6-4 is designed with the following format in mind. Starting from the left, the first column indicates the rank order of variation and is based upon the number of places that the item varies in rank as assigned by the individual regions. Here the items are arranged from the least to the most variable. Column two contains the written description of the potential impact, while column three shows the number of places that the item varied among the responses obtained from the participating regional representatives. This place variation is further described in the following column that shows the range over which the rank of the item varied. For instance, in table 6-3, the least variable item was the potential temporary harmful aesthetic effect of highway construction. This particular item varied only three places in rank, with the range of this variation being from twenty-eighth to thirtieth in importance.

The final column is broken down into three subheadings that show the number of times the impact factor appeared in each of three arbitrary groupings. The first subheading indicates the number of times that the item ranked between the first and the tenth most important in the hierarchy in a particular region. Items falling in this category are titled "Relatively Important." The next subheading indicates the number of times that the impact factor was between the eleventh and the twentieth most important item in a particular region. Items which are found in this particular grouping are assigned the title of "Conditionally Important." This means that their rank within the value structure of a region is conditional upon the appearance and importance attached to the other potential impacts. This particular group can be looked upon as a residual where all those items which are not important or unimportant are to be found. The third subheading shows the number of times that a particular impact factor was rated between the twenty-first and the thirtieth most important item within a given region. Items which are classified in this subheading are titled "Relatively Unimportant." Again, it must be emphasized that these relative rankings are derived from the numerical weights which are tabulated and presented in sample form in chapter 5.

To further clarify the importance of this final column, let us continue with the example of the place variation for the temporary aesthetic effects found in table 6-3. This particular impact was not rated in the top ten in importance by any of the participating regions, nor was it ever considered in the conditional group. All of the rankings assigned to this particular item indicate that it belongs in the category of impacts that can be considered relatively unimportant. This would indicate that the values associated with this particular impact are relatively less essential when comparison is made to the entire spectrum of potential effects.

The row items are classified according to the number of places that they were observed to vary. This classification is based upon a division into three arbitrary groups, which gives an indication of the extent of stability to be expected from the value associated with the item as the potential impact moves among regions. The first group contains those impacts that are regarded as "stable" in nature, since their number of places varies ten or less between the high and low rank extremes. The stable nature of the observed ranking would tend to indicate that highway planners could count upon a uniform level of importance being attached to a particular item as potential projects move between impacted regions. This particular level of importance depends primarily upon the worth of the item itself and is least affected by external forces that can significantly alter the assigned weights.

The second set of impacts are those that vary between eleven and fifteen places in their ranking. Items in this category are classified as "Conditional," since their variation tends to depend upon the current circumstances within the region, which partially affect the present importance attached to the item. Such surrounding conditions might include elements of the existing political or economic climate, as well as the general character of the area itself. This character would be reflected by whether the region was urban or rural, whether it was industrial or white-collar oriented, and certain residential traits such as personal income, age, and educational background. All of these surrounding conditions could have various effects upon the ranks assigned to particular potential impacts.

Table 6-3
Analysis of the Place Variation of the Detrimental Items Among Thirteen Planning Areas

Rank	Item	Number of Places Varying	Range	The number of times the impact factor appears in each group: Relatively Important 1-10	Conditionally Important 11-20	Relatively Unimportant 21-30
1	Temporary Aesthetic Effects	3	28-30	0	0	13
2	Property Values	6	1-6	13	0	0
3	Satisfaction with Government	6	25-30	0	0	13
4	Health Effects of Pollution	6	1-6	13	0	0
5	Access to Emergency Facilities	7	1-7	13	0	0
6	Aesthetic Value of the Right-of-Way	9	10-18	2	11	0
7	Number of Jobs	10	1-10	13	0	0
8	Personal or Business Contacts,	10	19-28	0	2	11
9	Barrier Effect	10	17-26	0	2	11

Stable

10	Personal or Group Stress	11	10-20	2	11	0	Conditional
11	Blend Highway into Background	12	10-21	3	9	1	
12	National Defense	12	17-28	0	4	9	
13	Visual Quality of Highway	13	12-24	0	9	4	
14	Temporary Economic Effects	13	17-29	0	2	11	
15	Pattern of Land Development	14	4-17	9	4	0	
16	Number of Housing Units	14	7-20	7	6	0	
17	Safety on Adjacent Highway	14	2-15	12	1	0	
18	Aesthetic Effects of Pollution	15	3-17	9	4	0	
19	Public Participation in Government	15	15-29	0	2	11	
20	Amount of Open Space	15	2-16	8	5	0	
21	Neighborhood Stability	15	14-28	0	6	7	
22	Level of Income	16	1-16	11	2	0	Volatile
23	Community Security	16	12-27	1	3	9	
24	Municipal Services	17	10-26	1	9	3	
25	Financial Capability of Government	17	9-25	1	7	5	
26	Community Cohesiveness	17	6-22	1	8	4	
27	Community Oriented Contacts	18	11-28	0	7	6	
28	Number of Historic Sites	19	9-28	1	3	9	
29	Number of Welfare Recipients	20	7-26	2	7	4	
30	Number of Business Firms	22	1-22	8	4	1	

Table 6-4
Analysis of the Place Variation of the Beneficial Items Among the Thirteen Planning Areas

Rank	Item	Number of Places Varying	Range	The number of times the impact factor appears in each group: Relatively Important 1-10	Conditionally Important 11-20	Relatively Unimportant 21-30
1	Temporary Aesthetic Effects	3	28-30	0	0	13
2	Property Values	6	3-8	13	0	0
3	Health Effects of Pollution	6	1-6	13	0	0
4	Personal or Business Contacts	6	22-27	0	0	13
5	Satisfaction with Government	8	23-30	0	0	13
6	Level of Income	9	1-9	13	0	0
7	Access to Emergency Facilities	9	3-11	12	1	0
8	Number of Jobs	10	1-10	13	0	0

Stable

9	Temporary Economic Effects	11	19-29	0	2	11	Conditional
10	Personal or Group Stress	11	9-19	5	8	0	
11	Community Cohesiveness	11	10-20	1	12	0	
12	Community Oriented Contacts	11	17-27	0	7	6	
13	National Defense	12	17-28	0	2	11	
14	Aesthetic Effects of Pollution	13	5-17	7	6	0	
15	Number of Business Firms	13	1-13	8	5	0	
16	Blend Highway into Background	14	12-25	0	10	3	
17	Aesthetic Value of the Right-of-Way	14	9-22	4	8	1	
18	Safety on Adjacent Highway	14	2-15	10	3	0	
19	Barrier Effect	14	16-29	0	3	10	
20	Amount of Open Space	15	2-16	9	4	0	
21	Visual Quality of Highway	16	12-27	0	8	5	Volatile
22	Public Participation in Government	16	14-29	0	2	11	
23	Number of Housing Units	17	7-23	3	8	2	
24	Number of Historic Sites	17	13-29	0	4	9	
25	Neighborhood Stability	17	12-28	0	6	7	
26	Financial Capability of Government	18	7-24	1	9	3	
27	Pattern of Land Development	18	1-18	12	1	0	
28	Community Security	20	8-27	1	3	9	
29	Number of Welfare Recipients	22	5-26	4	8	1	
30	Municipal Services	23	4-26	1	10	2	

The third category contains those items which varied sixteen or more places in their relative ranking. These items are labeled as "volatile" in their weights and, consequently, are extremely unpredictable for planning purposes as a corridor moves from one region to another. These items might derive their volatile nature from several regional characteristics. First, there might be a highly emotional aspect associated with a particular item in a given area. This would occur where the item was of vital personal or community concern so that its existence was made known to a majority of the participating respondents. Secondly, this extensive variation might reflect strong local character traits that tend to place emphasis upon one or more aspects of the community effect of highway construction. Thirdly, in addition to being emotional in character, the issue may well have a topical, local, or national application which reinforces the importance of the item within a specific region. In evaluating volatile impacts, highway planners would do well to direct a sizeable amount of their investigative work on these items, if their existence can be anticipated as a result of highway construction.

At this point, it is worthwhile to further emphasize the importance attached to the coordinate classification structure, which is being applied to the analysis of the thirty impact items. To achieve this reinforcement, the general format is reproduced in table 6-5. The columns in this table again show the relative importance of the particular item. A given impact is classified as important, conditional, or unimportant in its grouping according to where it appears a majority of the time. The row categories classify the items according to their extent of variation, with the three alternatives being stable, conditional, and volatile. Care

Table 6-5
Cross-Classification by the Characteristics of Ranking

	Classification by the Importance Group of Majority Appearance		
Classification by the Extent of Variation in Rank	Relatively Important 1-10	Conditionally Important 11-20	Relatively Unimportant 21-30
Stable Variation (1-10 places)	0	Z	0
Conditional Variation (11-15 places)	Z	Z	Z
Volatile Variation (16-30 places)	X	Z	X

Note:

O = Most reliable for planning purposes

X = Most strongly influenced by regional character

Z = Predictability for planning purposes is dependent upon relative importance of other items and surrounding conditions.

should be taken that the meaning of the two conditional entries is fully understood. In the *column* category, conditionally important refers to the relative position assigned to a particular item as it is compared to other items contained in the two extreme categories. The conditional entry in the *row* series refers to regional characteristics or circumstances that surround and influence the extent of variation.

In emphasizing the importance of this cross-classification, it should be pointed out that there are three sets of coordinates that deserve particular attention from highway planners. The first is indicated by zeros and refers to those items that are stable in variation and either important or unimportant in their grouping. Items in these cross-classifications can be viewed by highway officials as containing the highest degree of predictability for planning purposes. On the other hand, the impacts in the coordinates marked by an X are those effects that are both volatile in their variation and either relatively important or unimportant. These appear to contain the impacts that are the most deceptive from a planning standpoint. The concern attached to these items can be strongly affected on occasion by the characteristics of the region or the emotional nature of the item within the particular area. Consequently, planners, who are unaware of the potentially volatile nature of the item, may feel the importance of the item can be relied upon to a far greater degree than is actually justified. As a result, where variations occur they can be the source of severe conflict and misunderstanding within the decision-making process. In recognition of this point, these particular effects should be given conscious attention by highway officials, with their appearance in specific areas possessing the potential of generating significant amounts of either opposition to or demand for the construction of the proposed facility.

The third set of coordinates marked by a Z expresses various degrees of uncertainty. Those potential impacts that contain an element of conditionality are to be reviewed with skepticism, since their value depends, at least in part, on factors which are external to the impact itself. As a result, little can be said about these items except that they can be expected to vary in some functional relationship with the observable characteristics of the region affected.

Returning to tables 6-3 and 6-4, there is a tendency for this system of analysis and classification to be affected by the extreme rankings that might be assigned by one or two regions. For example, although an item may be rated important in eleven out of thirteen regions, it can appear much more volatile than it really is if the remaining two regions rated the impact in the relatively unimportant range. This would allow the item to have upwards of a thirty place variation in its rank. Consequently, in order to correct for this possible deception, if an item is contained in one of the three groupings eleven or more times, then some verbal notations will be made in the following analysis that the extent of variation may not be as severe as it appears. Although this process is arbitrary, it allows for some justifiable corrections, where particular regional traits may be so overriding that the item is assigned a level of importance which places it at odds with the majority of participating groups.

Analytical Profile of the Detrimental Place Variation

Table 6-3 contains an analysis of the place variation of the detrimental items among the thirteen planning regions. Following the format outlined above, the first nine items are classified as stable, since the extent of their place variation is ten or less. Within this stable category four of the entries were considered relatively important, four were considered relatively unimportant, and one was conditionally important relative to the appearance of other items in the questionnaire. Considering this breakdown, it should be immediately apparent that the respondents in all regions were both consistent in the way they regarded certain items as well as being able to purposefully separate those impacts which were of a vital nature from those which were, in comparison, not as crucial to the characteristics of the area.

The four items that appeared "Stable-Important" were highly personalized and reflected the potential harmful effect of a highway on the individual within the affected community. These items included the effect upon property values, the health effects of pollution, the accessibility to emergency facilities, and the number of jobs. Each of these personal items apparently possesses an element with which the respondent could personally identify. On the other hand, the "Stable-Unimportant" items are less personalized and refer more to the quality of life within the community. These are the temporary aesthetic effects, the satisfaction with government, and the barrier effect of highway construction. An exception would appear to be the effect upon personal or business contacts, which possesses an individual side. Since the respondents were asked to reply using their town of residence as a frame of reference, the appearance of this item in this particular coordinate position is in part a reflection of the separation of the town of residence from the town of employment on the part of the average respondent. Clearly, these four items can be counted upon as generating relatively lesser public concern in the location process, especially where the residents of the affected area correspond most closely to the characteristics of the sample group.

The next set of eleven items contains those impacts whose importance is conditional upon the current circumstances that exist in the test region. As noted above, such surrounding conditions as the current economic climate, the present level of social stability, and current local issues of topical interest may well have been the cause of the more extensive variation observed in the rankings assigned to a particular item. Of the eleven items, five could be considered "Conditional-Important." These would include the pattern of land development, the number of housing units, safety on adjacent highways, the aesthetic effects of pollution, and the amount of open space. All the items except safety are physical and relate to the impact that a proposed highway can have on the tangible environment which surrounds the region.

Three of the items can be considered "Conditional-Conditional" including personal or group stress, blending the highway into the background, and the visual quality of the highway. The latter two are design aspects and their appearance in this coordinate indicates that, while their existence is important, they are flexible and dependent upon other aspects of the community involved. Finally, four of the items could be considered "Conditional-Unimportant." These included the effect upon national defense, the temporary economic effects, the effect upon public participation in government, and the effect upon neighborhood stability. All of these were rather strongly unimportant except the last, which had several entries contained in the "Conditional-Conditional" coordinate. The ambiguity attached to this item might be expected from a group of regions that ranged between rural and urban in nature.

Generally, the items in the conditional variation category refer to either the physical or aesthetic aspects of the affected community. The major exceptions to this observation are three semipersonal items, and one that refers to the relationship between individuals and the government. Each of these entries, however, is subject to the correction factor noted above, since the majority of responses are located heavily in one of the three importance categories. Consequently, the extent of place variation is strongly affected by the extreme responses obtained from only one or two of the participating regions. Each of these items appears to be highly stable in its response pattern with safety on adjacent highways tending to be more "Stable-Important" with personal or group stress tending to be "Stable-Conditional," and both temporary economic effects and public participation in government tending to be "Stable-Unimportant."

The final variation grouping contains nine items that are highly volatile. Each of these items varied at least sixteen places resulting in most of them being considered at least once in all of the three importance categories. This grouping includes the level of income and the number of business firms, which could be considered "Volatile-Important." The effect upon municipal services, the effect upon the financial capabilities of government, community cohesiveness, community-oriented contacts, and the number of welfare recipients fall into the "Volatile-Conditional" coordinate. And finally, community security and historic sites which could be considered "Volatile-Unimportant." Clearly, a majority of the items refer to the community as a whole with the extent of their variation being partially attributed to the emotions that momentarily surround these community oriented elements. When threatened by disturbance from external forces, it could be expected that these items would serve as significant rallying points for the development of opposition to proposed highway plans. Most of these items were not consciously valued by the respondents unless there was some element already existing in the community which had brought the importance of these items to the attention of the participating group. However, where this element exists, it is clear that the participants were able to single out the item for particular attention and to give it extra weight within the decision-making flow.

Only the effect upon the level of income could be considered more stable than it initially appears. Within inferring more knowledge than is available, it would appear to be safe to say that this item could be considered at least conditional in variation and important in its value to the community.

Analysis of the Beneficial Place Variation

Table 6-4 applies the same analytical format to a review of the regional rankings assigned to the beneficial effect of the thirty factors. The first category contains eight items that can be considered stable in their place variation. Within the category, five of the items are "Stable-Important." These items include the effect upon property values, the health aspects of pollution, the effect upon the level of income, the accessibility to emergency facilities, and the effect upon the number of jobs. As with the detrimental side, each of these items are personal in nature with the only new addition being the effect upon the level of income, which moved up from the "Volatile-Important" category on the detrimental side. The other items all appeared stable and important on the detrimental side as well.

The remaining three items, the temporary aesthetic effects, the effect upon personal or business contacts, and the effect upon satisfaction with government, are rated "Stable-Unimportant" by the classification system. These three items also received the same classification on the detrimental side indicating that they too could be counted upon in highway planning. One element which should be emphasized, however, is that the coordinate position assigned to the effect upon personal or business contacts may be affected by two peculiarities. First, the sample group was suburban in nature, which tended to make their town of residence separate from their town of employment. Consequently, the respondents for the most part could not visualize a highway passing through their town of residence which would also affect their business contacts. Second, regarding the personal aspects of the item, the sample group is a highly mobile middle-class collection of participants who might not place the same emphasis upon personal contacts that would be found in more urbanized areas. If existing contacts are broken, the mobility factor would allow the respondents to establish new lines of communication.

The next category contains twelve items that are conditional in their variation, since they depend in part upon existing circumstances in the community. Because several community oriented impacts appear, this category is not as physically or aesthetically oriented as it was on the detrimental side. Four of the items can be considered "Conditional-Important," including the aesthetic effects of pollution, the effect upon the number of business firms, the safety on adjacent highways, and the amount of open space. Five of the items are classified "Conditional-Conditional" including personal or group stress, community cohesiveness, community oriented contacts, blending the highway into the background, and the aesthetic value of the right-of-way. Finally, three of the items can be considered "Conditional-Unimportant." This would include the temporary economic effects, the effect upon national defense, and the barrier effect of highway construction.

Two additional characteristics are associated with the conditional category. First, it contains three items that are apparently assigned by the extreme rankings of one or two planning regions. This would include the temporary economic effects, the effect upon community cohesiveness, and the effect upon national defense. Each of these items can be considered more stable than it would initially appear when only considering its range of variation. Secondly, seven of the items also appear in the conditional category on the detrimental side. This includes temporary economic effects, the effect upon personal or group stress, the effect upon national defense, the aesthetic effects of pollution, blending the highway into the background, safety on adjacent highways, and the effect on open space. The majority of the additions to the conditional category involve increased stability on the part of items that were volatile on the detrimental side.

Finally, the volatile category contains nine items that showed extreme elements of variation as witnessed by their fluctuation in rank. Of these items, one, the pattern of land development, can be considered "Volatile-Important," and five can be considered "Volatile-Conditional," including the visual quality of the highway, the number of housing units, the financial capability of government, the number of welfare recipients, and the effect upon municipal services. The last classification contains those items that are "Volatile-Unimportant," with this group including the effect upon the public's participation in government, the number of historic sites, neighborhood stability, and community security. In this volatile category, two items can be considered as having greater stability than would be initially apparent. These are the effect upon public participation in government, and the effect upon the pattern of land development. Only five of the items appeared in this volatile category on both the beneficial and detrimental side with a significant amount of interchange apparently having taken place between the volatile and conditional categories as the respondents moved from one side of the ledger to the other.

Review of Operational Hypotheses

It would appear from the analysis of place variation shown above that some preliminary comments can be offered regarding the substantiation of the operational hypotheses stated in section 4.4. In particular, several of the hypotheses could be considered as verified in full or in part. The first hypothesis concerned the physical ability of the participants to perform the operation and develop a set of relative impact weights. That this operation can be performed by representatives of the public is evidenced by the 200 valid responses which were obtained at various regional planning meetings. For the most part, the people were not confused by the mechanical operation and were willing to cooperate with a project that they felt would lead to substantial benefit for their particular region. The point is recognized that the participants can be considered above average in their ability to perform this operation. However, the hypothesis regarding the public's ability to participate in such an activity seems in part established. The remaining task would be to simplify the techniques, so that the questionnaire could be administered on a random sample basis, if this particular method is chosen as the future selection procedure to follow.

The second hypothesis offered was that participants were sufficiently aware of their own community attitudes to articulate these in a way which would facilitate highway route location decision making. Again, it appears from the analysis of place variations that this hypothesis has been partly established. This is especially true of those items that were stable in variation and which were assigned to one or more of the importance categories consistently by the participants. If some set of values were not the underlying factors that governed the pattern of selection, the responses would demonstrate considerably less consistency and be more randomly located among the coordinate positions than they have been observed to be. The fact that so many of the items are contained in the conditional or volatile categories indicates that the second half of this particular hypothesis is also plausible in that, not only do these weights tend to change over time, but they also vary with the supporting conditions that exist within the individual regions.

The third hypothesis stated that significant variations in the weights assigned to the particular items were to be expected between the participating regions. This particular hypothesis is again substantiated by the extent of place variation. This is most fully revealed by the fact that some of the items vary upwards of twenty places between the extreme high and low rank assigned by various regions. The extent of variation is often accounted for by recognizable area characteristics in the sections outlining the regional profiles. Consequently, the evidence of variation is significant in that it lends support to the plausibility of the hypothesis. Furthermore, the analytical profile that accompanies the presentation of each set of regional weights establishes that not only is this variation related to recognizable regional characteristics, but that these characteristics extend down to the town or neighborhood level as well.

The fourth hypothesis stated that the respondents possessed the capability to feel differently regarding the potential beneficial as opposed to detrimental effect as it might appear with respect to the same item. As evidence that this hypothesis has also been established, reference may be made to tables 6-1 and 6-2 that list the order in which particular potential impacts were ranked. In no instance was the rank order identical for the beneficial and detrimental sides and, in many cases, there was substantial variations in rank assigned to a particular time on the beneficial as opposed to the detrimental side. This was especially true with those items that were conditional in nature in both their place variation and in their ranking. Such items, along with those contained in the volatile category, are related to both current circumstances and emotional conditions existing within the test region. Consequently, the importance of the item varies substantially within a region from the rank assigned on one side as opposed to the other in the weighting process.

The fifth and final hypothesis concerned the ability of the planners to review the results and to offer comments regarding their validity within the appropriate region. Substantiation of this hypothesis can only be hinted at through the quality of the testimony offered by the planning director as it relates to the analytical profile of this particular region. However, it should be noted that the plausibility of the four previous hypotheses offers the planner the opportunity for analysis based upon regional characteristics, since the extent of variation itself was significant enough to facilitate this type of review.

Review of the Presentation of the Weights and the Analytical Format

The following sixteen sections contain individual presentations of the numerical weights tabulated for each of the participating groups. This presentation is in tabular form, and is accompanied by an analytical profile that attempts to relate both the relative weight and the rank order in which the item appears to the known characteristics of the participating group. The tabular presentation assumes the following format:

Item Rank	Detrimental	Category	Weight	Beneficial	Category	Weight

Starting from the far left, the first column repeats the rank associated with the particular potential impact as shown in tables 6-1 and 6-2. The second lists the detrimental item that assumes that rank within the region. The third column indicates the category in the questionnaire where the particular item appeared. The number refers to the numerical group where the item was found, while the latter refers to the position of the item itself. The next column contains the numerical weight associated with the potential impact. This weight is adjusted for both the relative position of the item within the category and the relative position of the category among all six major classifications. The item is not adjusted for the weights attached to the overall user as opposed to the nonuser effects of highway construction. Adjusting for this factor would not change either the relative position of the item nor the relative differences which exist between one numerical weight and another. Consequently, in order to avoid any confusion with the interpretation of the results through the addition of more than four decimal places, this additional weight was omitted from the tabulation.

In order to properly interpret the implications of the weights, let us assume that there are two potential impacts *A* and *B*. Within the participating group, *A* might be assigned a weight of .05, while *B* might be assigned a weight of .01. The meaning of this relative difference is that the quasi-market impact associated with item *A* is to be considered five times more important than the quasi-market impact associated with item *B*. The final three columns repeat the presentation of items, questionnaire category and numerical weight, but from the beneficial as opposed to the detrimental side.

The above format is continued until all of the thirty potential impacts have been ranked and weighted. After this has been accomplished, two additional pieces of information are also provided. The first shows the relative weights associated with the six categories that were used to adjust the ratings assigned to the items within categories and which permit the investigator to compare all thirty items simultaneously. The weights assigned to these six categories are indicated by having the first column change its heading from "Item Rank" to "Category Rank." The second piece of information shows the relative weight assigned to the user as opposed to the nonuser effects of highway construction. Again, to indicate the appearance of this category, the heading of the first column changes from "Category Rank" to "General Rank."

6.2 Analytical Profile of the Aggregate State Weights Assigned by Citizen Participation

The following profile provides an analysis of the weights that were developed for the state as a whole based upon the answers supplied by the individual regional participants. The thirteen aggregate regional answers were adjusted by the projected percentage of 1970 state population that each of the participating regions would possess.[a] This population weight was used as a common denominator by which the regional answers could be summed to yield an estimate of the value structure for the state as a whole. The number of valid responses utilized in this aggregation process totaled 201. From this set of population-weighted responses, the following statewide weights were developed for highway corridor location purposes.

[a]The population projections were supplied by the Connecticut Bureau of Highways and were based upon "Connecticut Interregional Planning Program Population Projections," Hartford, Connecticut, published in December 1968.

In general, the respondents were heavily concerned with the personal effects of new highway proposals as shown by the emphasis placed upon the items found in both the health and safety and the economic categories. The element of personal concern was strongly modified by the attention devoted to several of the environmental items in the land use and aesthetic categories. This particular response pattern appears on the beneficial and detrimental sides, since the first eleven items are the same in both groupings. The importance of these eleven items is reinforced by the analysis that accompanied tables 6-3 and 6-4. In those tables, ten of the items were shown to be relatively important by a majority of the participating regional representatives. The only item of a marginal nature in this cross classification scheme was the effect upon the number of housing units, which appeared relatively important on the detrimental side, but only received conditional importance on the beneficial side. In the present analysis, this particular item received the eleventh rank, which again indicates its marginal relationship in the relatively important category.

The next set of items (12-20) focuses upon specific impacts in the aesthetic, social, and political categories. There is some tendency to stress the design aspects in the aesthetic category, and those political items that refer to the quality of the physical relationship between the local governing body and the residents of the region. These impacts, for the most part, were shown to be conditionally important in tables 6-3 and 6-4, with their ranking being a strong function of the number and importance of the other potential impacts, which were offered on the questionnaire.

The final set of ten items also contains a great number of potential impacts that can be considered personal or social in nature. However, as opposed to the personal items that appeared relatively important, it is apparent that these latter items are recognized as being far less vital to the highway planning process. In general, they refer either to the quality of the social environment that surrounds the community or the quality of the nonphysical relationship which exists between the residents and local government. Other impacts that also appear in this last grouping are the items which were deemed least important in the economic and health and safety categories. As with the previous two sets, the relative rankings of these items were, in part, predicted by the analysis of place variation. Here, all but two of the items were shown to be consistently unimportant on both the beneficial and detrimental sides. The two remaining impacts were again items whose fluctuation between groupings indicated the marginal nature of their association with the preceding importance category.

The results of this state aggregation must be modified somewhat by noting that there has been a bias introduced into the sample group. In particular, the majority of the regional participants were white, middle class, suburban property owners, whose interests would lie more in the direction of maintaining their physical and economic condition by protecting the status quo from the external forces of change. This group was supplemented in most regions by additional

Table 6-6
Weighted Coefficient and Order of Rank for Each of the Thirty Potential Impacts (Group: State-Aggregate)

Item Rank	Detrimental	Category	Weight	Beneficial	Category	Weight
1	Health Effects of Pollution	6-C	.0495	Health Effects of Pollution	6-C	.0502
2	Property Values	3-C	.0471	Number of Jobs	3-A	.0496
3	Accessibility of Emergency Facilities	6-A	.0462	Level of Income	3-E	.0468
4	Number of Jobs	3-A	.0460	Accessibility of Emergency Facilities	6-A	.0456
5	Safety on Adjacent Highway	6-E	.0441	Pattern of Land Development	5-A	.0445
6	Level of Income	3-E	.0432	Safety on Adjacent Highway	6-E	.0434
7	Pattern of Land Development	5-A	.0394	Property Values	3-C	.0428
8	Amount of Open Space	5-C	.0393	Amount of Open Space	5-C	.0425
9	Aesthetic Aspects of Pollution	2-E	.0383	Number of Business Firms	5-B	.0385
10	Number of Business Firms	5-B	.0374	Aesthetic Aspects of Pollution	2-E	.0374
11	Number of Houses	5-D	.0362	Number of Houses	5-D	.03646
12	Blend Highway into Background	2-B	.0345	Personal or Group Stress	6-D	.03642
13	Aesthetic Value of Right-of-Way	2-C	.03368	Aesthetic Value of Right-of-Way	2-C	.0338
14	Personal or Group Stress	6-D	.03365	Number of Welfare Recipients	3-B	.0330
15	Visual Quality of Highway	2-A	.0319	Community Cohesiveness	7-A	.0321
16	Community Cohesiveness	7-A	.0318	Blend Highway into Background	2-B	.0316
17	Municipal Services	4-A	.0315	Municipal Services	4-A	.0308
18	Number of Welfare Recipients	3-B	.0313	Financial Capability of Government	4-C	.0300
19	Financial Capability of Government	4-C	.0312	Visual Quality of Highway	2-A	.0286
20	Number of Historic Sites	5-E	.0296	Community Oriented Contacts	7-E	.0285
21	Community Oriented Contacts	7-E	.0285	Neighborhood Stability	7-C	.0281
22	Community Security	4-D	.0283	Barrier Effect	7-D	.0268
23	Neighborhood Stability	7-C	.0280	Community Security	4-D	.0259
24	Barrier Effect	7-D	.0269	Personal or Business Contacts	7-B	.0254

Table 6-6 (Cont.)

Item Rank	Detrimental	Category	Weight	Beneficial	Category	Weight
25	Personal or Business Contacts	7-B	.0254	Participation in Government	4-B	.0240
26	Temporary Economic Effects	3-D	.0249	Number of Historic Sites	5-E	.0239
27	Participation in Government	4-B	.0246	National Defense	6-B	.0233
28	National Defense	6-B	.0242	Temporary Economic Effects	3-D	.0216
29	Satisfaction with Government	4-E	.0195	Satisfaction with Government	4-D	.0183
30	Temporary Aesthetic Effects	2-D	.0216	Temporary Aesthetic Effects	2-D	.0162
Category Rank	**Detrimental**	**Category**	**Weight**	**Beneficial**	**Category**	**Weight**
1	Health and Safety	8-E	.1978	Health and Safety	8-E	.1991
2	Economics	8-B	.1927	Economics	8-B	.1940
3	Land Use	8-D	.1820	Land Use	8-D	.1884
4	Aesthetics	8-A	.1511	Aesthetics	8-A	.1478
5	Social-Psychological	8-F	.1408	Social-Psychological	8-F	.1411
6	Political	8-C	.1353	Political	8-C	.1292
General Rank	**Detrimental**	**Category**	**Weight**	**Beneficial**	**Category**	**Weight**
1	Nonuser Effects	1-B	.5494	User Effects	1-A	.5154
2	User Effects	1-A	.4505	Nonuser Effects	1-B	.4845

participants, who were asked to participate in the questionnaire by the regional planning director as a compensating force to offset the lesser attention devoted to the needs and concerns of urban residents. Even with this correction factor, however, the social concerns, which are more vital to the residents of an urban core area, are consistently awarded lower ranks by the participants. As a result, some reservation should be introduced as these weights and the accompanying analysis are applied to urban highway proposals.

The purpose of this state aggregate and the accompanying analysis is twofold. First, the state weighting system provides a broad description of the statewide value structure. It points out those factors which are relevant in the aggregate for state highway planning and provides a set of state characteristics which help to account for the observed rankings. Secondly, it provides a standard based upon public participation for comparison with the other sets of weights obtained from the alternate participating groups. In particular, it should point out key variations in the value structures associated with either the regional planning directors or the highway officials as opposed to individual state residents. Such variation should indicate areas where adjustments will have to be made in order to bring the participating groups to closer agreement. Such adjustments can take the form of either alterations in existing planning procedures or an increase in the information distributed to the public, which informs it of the forces that lie behind the current planning process.

In addition, the existence of a state aggregate weighting formula points out the possible sources of conflict that can appear between state and regional transport values. In certain cases, corridors that are undesirable from a local or regional point of view, are necessary when considered from the perspective of the state as a whole. Consequently, the evaluation of a set of statewide weights relative to regional considerations can point out possible areas of difficulty, which then may receive additional attention from highway planners.

Analysis of the Detrimental Results

On the detrimental side, the most important item to the participants as a group was the health effects of pollution. This ranking appears to be the result of two separate forces. First, in many areas throughout the state there is a genuine threat to individual health caused by the growing level of environmental pollution. This is especially true in the urban or industrial areas, with regions containing major rivers expressing an added level of concern. Clearly, the respondents were able to separate the health and aesthetic aspects of pollution, indicating that the responses were not an emotional reaction to the current national interest directed toward the level of pollution and environmental quality. The respondents appear to be indicating that they recognize the relationship between regional pollution and highway construction in terms of both the waste contaminants distributed by passing motor vehicles and the undesirable industrial pollution which results from the developmental effects that often accompany highway construction. Obviously, the respondents are not willing to accept highway construction where the result will be a noticeable increase in the risks associated with personal health.

The second source of concern with the health aspects of pollution originates from the intense interest that has been directed towards this topic on a national level. This is an item that has attracted considerable attention and publicity. The attention exists to the extent that a majority of the nation's population is aware of potential threat that this impact holds for the quality of the environment. As a result, people are strongly motivated by a national as well as a local concern with this particular item.

Other health and safety impacts were also considered important, with the accessibility of emergency facilities and the safety on adjacent highways being ranked third and fifth in the hierarchy. These two items reinforce the hypothesis that those impacts possessing a highly personal dimension which can substantially threaten the individual will receive direct attention from the respondents. The personal aspect of these two items is clearly evident and, as a result, the respondents can easily identify with the anticipated situation.

The safety on adjacent highways, in most cases, refers to the added traffic flow that will be directed toward existing secondary and urban roads if the proposed project is constructed. In many cases, these roads are inadequate and unable to handle the increased vehicular movement in a way that allows for a margin of safety, which is desirable from the point of view of those who must travel these routes. On the other hand, the accessibility of emergency facilities is vital in times of crisis, with a limited access highway at least presenting the potential for disrupting the flow of emergency vehicles. The respondents are indicating by the weight attached to this item that, where this potential exists, its impact should be considered quite strongly within the decision-making flow.

In comparing the ranks assigned to the health and safety items by the respondents as a whole to those assigned by the regional planners and the state highway officials, it is evident that there is a divergence in the level of concern. This divergence is evidenced by the tenth and thirteenth rankings assigned to the access to emergency facilities and the health effects of pollution by the highway officials. Even the planners felt that the access to emergency facilities should rank only eighth although they did feel that the health effects of pollution was the item of prime importance. This divergence on the part of the highway officials may be due to two forces. First, with respect to the access to emergency facilities, it may be that they regard this item in an impersonal way which allows them to take a more detached view of both the extent and meaning of variations in the vehicular movement during periods of crisis. They may see only minor changes in the pattern of movement for emergency vehicles which result in marginal increase from the personal point of view, the regional respondents feel that it should be given far greater attention than it now receives in the planning process. On the other hand the past experience of highway planners may lead them to feel that they presently direct the proper amount of attention to this problem.

Regarding the health effects of pollution, there is apparently a lag between the growth in importance of this item on the part of area residents and the recognition of this importance by its inclusion within the corridor location process. Lags of this type also appear with respect to items that were important to the public in the past, but whose position within the relative ranking scheme has diminished with the growth of new items. The instances of these lags would seem to be an indication that highway planners must introduce changes within their planning procedures to more fully develop their route location process in light of public desires. With respect to the third item of safety on adjacent highways, the three participating groups were in extremely close harmony regarding the importance they attached to this impact. Here, the attitudes of the highway planners closely match public desires.

In the economic category, three items appear to generate a high level of concern from the participants as a whole as shown by the second, fourth, and sixth rankings assigned to the effect upon property values, the effect upon the number of jobs, and the effect upon the level of income. This ranking is, in part, a reflection of the personal nature of the items considered and may tend to be slightly distorted due to the suburban nature of the sample. Such distortion would appear most strongly with respect to property value as the majority of respondents were homeowners. Distortions might also appear in the rank assigned to the number of jobs since in several cases the separation of the town of employment from the town of residence would make the respondent less concerned about the effect that a highway would have upon his job if the town of residence were to serve as the site for potential construction. As a result, this item may be even more important in the value structure than is indicated by its present rank.

With respect to these two items the ranks assigned by the highway officials were in near complete agreement with the aggregate importance of the factor. There was, however, a strong tendency on the part of regional planners to downgrade the importance of the effect upon property values. This may be due to their feeling that too much attention may have already been devoted to the property effects of highway construction, to the extent that environmental or social effects have been neglected in the corridor selection process.

The effect upon the level of income was the item drawing the widest variation in rank being placed first by the state highway officials, fourteenth by the regional planners and sixth by the aggregate of participants. The high ranking by highway planners may have been accounted for by their concentration upon the economic spin-off effect of highway construction, while the lower ranking by regional planners may also be a reflection of their attitude that highway planning has concentrated too hard on the economic aspects. Whatever the reason, however, it is clear that both groups were wide of the mark in mirroring the importance of this item to the residents of the state.

Switching to the land use category, it is evident that the traditional level of concern directed towards the land use impacts of highway construction by highway planners is for the most part supported by the state aggregate. In particular, four land use items exist as strong modifiers for the dominant ranks assigned to the economic and health and safety impacts. They are the seventh, eighth, tenth, and eleventh rankings assigned to the pattern of land development, the amount of open space, the number of business firms, and the number of housing units.

Of particular interest is the fact that the two land use items that refer to the quality of physical environment appear ahead of the more physical or economically oriented land use effects of highways. It would appear that while the participants were desirous of achieving economic development effects from highway construction, they were not interested in achieving these at a cost of significant sacrifices in the quality of the surrounding physical environment. In a state with heavy concentrations of populations, both the pattern of land development and the amount of open space are vital for the future of quality growth within the area.

Each of the items in this land use category, except for the number of housing units, received slightly more attention from highway planners than they did from the respondents as a whole. However, the extent of variation does not seem to warrant a significant change in the land use aspect of the corridor location process. A major point of variation does appear, however, when the rankings of the state aggregate are compared to the regional planning directors. The latter group devotes considerably less weight to all of the land use items including the effect a highway has on the number of housing units. This is rather surprising, given the fact that almost all the planning directors expressed a deep verbal concern for this particular impact, as a result of the extensive shortage of housing in a majority of the planning regions. In part, the pattern of response of the planning directors with respect to these land use items may be caused by their feeling that in the past too much attention has been devoted to land use impacts of highway proposals. However, even if this point is true, it would appear from the tabulated results that this particular level of attention has, and still is, warranted by genuine public concern for these items.

The aesthetic effects of highway planning do attract considerable attention from the participants as a whole, although in most cases the individual items do not reach the level of importance indicated by the regional planners, while in certain other instances it exceeds the level of concern indicated by highway planners. The two groups of planners felt that blending the highway into the background was the aesthetic item deserving the most attention, while the state aggregate felt that the aesthetic effects of pollution was the most important aesthetic item, being ranked ninth. Clearly, they felt that the effect of pollution on the quality of the environment was an item to be considered quite heavily in the route location process. That official concern with the aesthetic effects of pollution was not as strong as in the state as a whole may be further evidence of the lag between the growth and importance of the item on the individual level and the recognition of this growth on the part of state planners.

There was also some initial feeling that highway planners may be lagging behind state residents in their recognition of a lesser importance attached to the design aspects of highway construction. However, at least as far as the results obtained in this questionnaire are concerned, the evidence is inconclusive. The only case where this might possibly be true is with respect to blending the highway into the background and, even here, the difference between the responses supplied by regional planners as opposed to the state aggregate demonstrates a far greater degree of variation than that of highway planners. Apparently, the design aspects, at least where the state aggregate is concerned, are conditional upon the appearance of other impacts within the affected area. Only the aesthetic effects of pollution have a strong enough appeal to appear important in its own right in the weighting scheme and, even here, some element of conditional evaluation is introduced by the extent of variation in the range of ranking over the thirteen regions.

In the political category, the respondents as a whole indicated a level of concern that was approximately on par with the rankings assigned by the state highway officials, with a slight deviation shown between state answers and those of the regional planners. With respect to municipal services, the state aggregate showed a greater degree of concern with this item than either of the planning groups. However, the seventeenth rank assigned to this item indicates either a lower level of public concern than would initially be anticipated, or a failure to see the effect that highway construction can have on the level of municipal services. In part, this may be due to the already low level or poor quality of services that exists in many of the planning regions. More concern was evidenced as the regional character appeared more urban in nature. However, even in these areas, the importance of municipal services rarely allowed it to escape the conditional grouping in importance.

With respect to the harmful effect that highway construction may have on the financial capabilities of government, the respondents, by ranking this item nineteenth, were overlooking the relationship between this item and the economic development of the area. If economic factors are injured by the construction of a highway this can potentially have a strong effect upon the tax base of the community. Consequently, unless the services provided by this tax revenue are considered unimportant, the financial effect of a highway upon local government should be considered much more important in the decision-making flow. Considering that the majority of local tax revenue is allocated to education purposes and the respondents in general were a property-owning group, the relatively low importance attached to this item is surprising and should deserve greater attention in highway planning. Curiously enough, highway planners felt that this item deserved even less attention than the regional respondents. This may be due to either the relative unimportance of the item itself or the feeling that highway construction in general does not exert a significant influence upon this particular item.

Finally, the items contained in the social psychological category were allocated in a majority of cases to the relatively unimportant category with the sole exception being the potential harmful effect on community cohesiveness, which ranked sixteenth. As noted earlier, the slightly suburban bias in the sample group may tend to partly account for this generally low level of public concern directed towards social linkages. This would be especially applicable where the rating of personal or business contacts were concerned, since there is a tendency to separate the town of residence from the town of employment. This minimizes the effect that highway construction in the town of residence can have on employment contacts as well as opportunities. The apparent unimportance of community-oriented contacts, on the other hand, may in part be attributed to the high mobility of the suburban resident who, if the social linkages are disrupted, can more easily establish them in a new area. Obviously, these concerns, along with neighborhood stability and the barrier effect of a highway, would be of far greater importance in an urban area, where such linkages contribute not only to the quality of the social environment but also to the ability to carry on daily activities within the limits of the urban core.

The responses of the highway planners were not on par with those received from the state as a whole. Variation is shown by the lower weight attached to community cohesiveness and the greater importance that is attached to community-oriented contacts. The latter item, along with the nineteenth rank assigned to the barrier effect by the highway planners, may in part be a recognition of the importance of these items in urban areas and the difficulties that have arisen in the past where highway planning has tended to neglect these particular issues. The weights assigned to these two impacts may be an example of a case where highway officials are far more sensitive to the needs of urban areas than are the members of the regional group. This would be a reversal of the direction of the lag noted earlier with the general public lagging behind the highway planners in appreciating the social implications of highway construction.

As would be expected, the regional planning directors placed far greater emphasis on the social psychological items than any of the other participating groups. In particular, two of the items, neighborhood stability and community cohesiveness, were considered to be relatively important, while only one item, personal or business contacts, was considered to be relatively unimportant. The placing of these items so high up in the ranking scheme may signal a recognition of the full impact upon the community of highway construction, and the anticipation of further unfavorable consequences if those in charge continue to neglect these issues. Here, professional training has made the regional planners more sensitive to these items than even the highway planners are.

The only social psychological item that received even conditional attention from the aggregate of participants was community cohesiveness which ranked sixteenth. This particular ranking may have been, in part, the result of the community-oriented structure from which the sample group was drawn. Most of the individual participants have been working for a considerable length of time on issues of a regional nature and they recognize the need for purely political subdivisions to unite and to act on those problems that extend beyond city or town borders.

Analysis of the Beneficial Results

As noted earlier, there is a significant element of symmetry between the results obtained on the beneficial and detrimental sides of the questionnaire. This symmetry extends to the point where the first eleven items are the same in both categories. The appearance of stability in the statewide value structure is important for two reasons. First, although certain regions may indicate that items of lesser statewide importance do command considerable attention in their area, there is a fair degree of unanimity attached to the set of items which are generally considered to be important throughout the state as a whole. Obviously, most of the items, no matter what region is being considered, take on an element of importance that should be strongly considered in the decision-making process. Clearly, where differences of opinion appear regarding regional as opposed to statewide corridor choices, the essence of the conflict can be traced to a local characteristic that causes a deviation from the statewide value structure regarding highway planning.

Secondly, the element of symmetry indicates that wherever highway proposals are anticipated, planners in general can count on certain items as being generally important to the impacted agents. Although this information still leaves planners a considerable distance from the actual point of corridor choice, it does tend to simplify the investigation by allowing them to focus attention upon those variables which, as a rule, command the most popular attention.

The impact to be considered most important, if highway construction can result in its improvement, is the health effects of pollution. The participants singled out this item on both sides of the ledger, indicating that public concern has grown to the point where considerable attention should be directed towards the potential effect of highways upon this aspect of the environment. The state does have its share of environmental quality problems which can, in part, be attributed to both the use of motor vehicles and the growth of industrial activity which has, in the past, been induced by road building. This first-hand experience with the pollution problem, combined with the current level of national attention directed towards this topic, emphasizes its importance in the minds of the participants. Whether this particular item will continue to be important in the future, given both greater knowledge regarding its dollar impact or the changes that it brings in the surrounding economic and social conditions, cannot at this time be forecast. However, it is safe to say, at least for the moment, that this item should attract far greater official attention than it has in the past.

Two other health and safety items were considered important by the public as shown by the fourth ranking assigned to the accessibility of emergency facilities and the sixth ranking assigned to safety on adjacent highways. As noted earlier, these are impacts that directly affect the individual. Consequently, the respondent, since he is able to place himself in a situation that allows him to fully visualize the meaning of these items, attaches substantial weight to their importance within the decision-making flow. Highway planners, in the past, have attached approximately the same level of importance to the safety items as the participants, with some tendency to lessen the importance assigned to the emergency access. With respect to the latter, a great effort directed towards improving the flow of emergency vehicles might serve as a favorable balancing point to offset other undesirable impacts that could accrue to the affected community.

Accompanying the above concern for the personal items in the health and safety category is the strong emphasis based upon the personal economic impacts. The ranking of such items as the number of jobs, the level of income, and the effect upon property values indicates that the respondents are both aware and desirous of the economic spin-off effects from highway construction. This point is well recognized by highway planners and the level of concern attached to these items by this group is approximately equal to that of the state as a whole.

Surprisingly, for both groups the impact upon property values falls in importance as the respondents moved from the beneficial to the detrimental side. Although the item still commands considerable attention being ranked seventh by the aggregate of participants, it is clear that the respondents are not as concerned with highways increasing the value of their homes as they are interested in maintaining and avoiding losses in the value of their physical property. As on the detrimental side, the overall importance attached to this item may, in part, be due to the suburban middle-class bias. However, except for the heavily concentrated urban core area, it would be safe to expect that the relationship between property values and highway construction should be given the attention it received here.

Continuing in the economic category, the second and third rankings assigned to jobs and income indicate that the respondents value quite highly the industrialization effects of the highways. This is further emphasized by the ninth rank assigned to the effect on the number of business firms. Except in unusual cases, highway planners should be able to count upon these beneficial impacts as important for decision-making within a community. Exceptions might arise in cases where the community was already experiencing a tight labor market, or where the impacted area was predominantly rural or residential with little or no interest in furthering the extent of regional development.

There is a land use modifier attached to the concern with economic development. This modifier is shown by the fifth rank assigned to the pattern of land development and the eighth rank assigned to the amount of open space. These two items indicate that the respondents are interested in maintaining the quality of their physical environment through the proper control of industrial development. Although industrialization is desirable, it is only slightly less important to preserve or increase the amount of available open space by the conscious use of existing land. The benefits derived from proper land control and development should, in general, be recognized within the decision-making process.

The importance of these two items is recognized by the highway officials who place even a greater emphasis upon the beneficial impacts. The principle point of interest, however, is the low value associated with these items by the regional planning directors. They appeared not to have properly considered the importance of these two items within the highway planning process.

Another item that in the land use category attracted considerable attention is the affect upon the number of housing units which was ranked eleventh. The problem of the present housing shortage is a statewide phenomenon, with a considerable need for the development of new living complexes that will allow for an upgrading of low income residential facilities and an increase in the availability of homes directed towards the middle class. The regional planning directors, who are familiar with the extent of this problem on a first-hand basis within their region, placed an even greater emphasis upon this than the statewide participants. However, the highway officials felt the item deserved less attention within the route location process which may, in part, be a reflection of their attitude that the Bureau of Highways does not have direct control over the construction or availability of housing in an affected region.

The next group of items that attracted attention from the respondents, were those factors located in the aesthetic category. One of these items, the aesthetic effects of pollution, found its way into the important category being ranked tenth in the hierarchy. This citation is in addition to the prime importance attached to the health effects of pollution. The distinction in rank between the two impacts indicates that the respondents are less than emotional regarding the word pollution and, as a result, they are fully capable of separating in importance those polluting impacts that are physically threatening from those which affect the senses only.

The other aesthetic items were design oriented and ranked thirteenth, sixteenth, and nineteenth in the ordering. Their location in this second group of ten would indicate that their importance varies with the appearance of other impacts in the affected area. As a result, while it is important to develop highways that add to the quality of the surrounding environment, the relative importance of these design items is a function of the other effects which can be anticipated within the area. With respect to design features, highway planners placed approximately the same emphasis upon these items as was shown in the state as a whole. Even the regional planning directors were fairly close at times in their assessment of the public importance attached to the design items. However, both groups were somewhat wide of the mark in assessing the importance attached to blending the highway into the background. While the regional planners placed far less emphasis upon the aesthetic value of the right-of-way than did the state aggregate. In conclusion, both supplemental groups allocated less concern to the undesirable aesthetic aspects of pollution. Here the highway officials again demonstrated the widest variation, which may offer further evidence regarding the lag hypothesis.

In the political category the respondents from the general public were only conditionally concerned with the effect that highway construction might have upon both the financial capabilities of government and municipal services. This is shown by the seventeenth and eighteenth rankings assigned to these particular items. The lack of concern is rather difficult to explain in light of the suburban property-owning bias introduced in the sample. Initially, it might be felt that these two items would be exceptionally important to the residents of an impacted area, if they could be significantly benefited by the highway construction. Both items are directly related to corridor location decisions, since the financial capability of local government is predominantly dependent upon the tax base, which in turn reflects both industrial and personal property values. Since both of these latter items receive considerable attention on the beneficial and detrimental sides, it should follow that their revenue aspects would also be of major concern. However, this point is not borne out. With respect to municipal services, these items too, are affected by highway building, since added mobility and industrialization will not only increase the tax base but also generate added population pressures that require an expansion in services. Even when the linkage is presented on a per capita basis, these points do not seem to carry much weight in the hierarchy of values as established by the participating regional representatives.

Again, the items in the social psychological category receive little or no overall attention from the participating groups. All of the items, except community cohesiveness and community-oriented contacts, fall within the relatively unimportant category, with the latter of the two being right on the margin between conditionally important and relatively unimportant. At least when middle-class residents in Connecticut are concerned, the value of social institutions in general should play a lesser role in the decision making flow than those impacts that affect either the economics of an area or the individual aspects of health and safety.

As noted several times, this lack of concern with social items may, in part, be due to the middle-class bias of the sample. This group may tend to feel socially secure with their extent of mobility allowing them to reestablish social and community linkages if they have been disturbed by highway construction. As a result, when in turn these linkages are benefited, the addition of new ties is not as important, since the area of opportunity is already quite large for the establishment of contacts to the extent desired.

The importance of this observation would tend to diminish as highway construction moved closer to the urban areas. Here the social needs of the urban resident are far greater, especially in ghetto areas where low mobility and ethnic constraints reduce the range of contacts available to the individual. Consequently, highway construction would possess greater value in these areas if it facilitated personal mobility and the development of linkages both within and between communities in the affected region.

The fifteenth rank assigned to community cohesiveness can, again, be partly attributed to the nature of the sample group. Most of the participants are members of regional planning commissions who, in serving in this capacity, recognize the value of the ability to solve problems which extend beyond city boundaries. As a result, they would place a higher weight upon highway construction which fostered a growth in intertown or interregional development programs.

In evaluating the social concern, relative to the importance attached to the items by the other participating groups, it would appear that the replies from the highway planners were about on par with the answers obtained from the state as a whole, except where the element of neighborhood stability was concerned. The highway planners appeared to place a far greater emphasis upon generating highway plans that improve the continuity of living patterns in recognizable residential areas. Emphasis was also placed upon this particular item by the regional planners.

In addition to this point, two other items are worth noting about the relative response patterns of the highway planners as opposed to the state as a whole. These revolve around the linkage items associated with the barrier effect of highway construction and the effect upon community oriented contacts. Clearly, there has been a growing concern, especially in urban areas with the effect of highway construction upon these two variables. Although none of the three groups tested directed more than conditional importance towards these two items, it is important to recognize that the development of highways can benefit as well as disrupt the pattern of social communication between and within recognizable communities. In steering highway development towards the improvement of these two impacts where possible, highway planners, in the future, may be facilitating the solution of social problems that presently strain the relationship between the urban center and the surrounding residential communities.

6.3 Analytical Profile of the Weights Assigned by the Participating Representatives from the Connecticut Bureau of Highways

The participating representatives from the Bureau of Highways were selected by the acting director of planning and the acting director of research. The criteria used to choose the respondents focused upon assembling a group whose views regarding the corridor route location process would provide an indication of the actual trade-offs presently applied by the bureau. Representatives were obtained from the divisions of planning, design, programming and scheduling, rights-of-way, and research. The number of valid responses obtained to the questionnaire totaled thirteen.

The pattern of response indicates a strong concern for both the economic and land use aspects of highway impact. This holds true of both the beneficial and detrimental sides and is consistent with the traditional attitudes of highway planners regarding the nonuser effects of highways upon the community. In particular, by relying upon the projection of future highway plans based upon estimated land use patterns and the resulting desired lines of travel, those responsible for planning clearly recognize the existence and importance of the interrelationship between highway construction and community land use effects. Furthermore, many of these land use impacts are highly economic in nature with

Table 6-7
Weighted Coefficient and Order of Rank for Each of the Thirty Potential Impacts (Group: Highway Officials)

Item Rank	Detrimental	Category	Weight	Beneficial	Category	Weight
1	Income	3-E	.0532	Income	3-E	.0507
2	Property Values	3-C	.0465	Pattern of Land Development	5-A	.0486
3	Jobs	3-A	.0440	Jobs	3-A	.0485
4	Pattern of Land Development	5-A	.0423	Open Spaces	5-C	.0433
5	Safety on Adjacent Highway	6-E	.0415	Property Values	3-C	.0404
6	Open Space	5-C	.0409	Welfare Recipients	3-B	.0401
7	Business Firms	5-B	.0400	Safety on Adjacent Highway	6-E	.0401
8	Welfare Recipients	3-B	.0398	Business Firms	5-B	.0392
9	Blend Highway into Background	2-B	.0385	Health Pollution	6-C	.0383
10	Accessibility of Emergency Facilities	6-A	.0384	Accessibility of Emergency Facilities	6-A	.0379
11	Housing	5-D	.0380	Neighborhood Stability	7-C	.0368
12	Aesthetic Value of Right-of-Way	2-C	.0377	Financial Capability of Government	4-C	.0354
13	Health Pollution	6-C	.0369	Blend Highway into Background	2-B	.0349
14	Community Oriented Contacts	7-E	.0356	Personal or Group Stress	6-D	.0347
15	Aesthetic Aspects of Pollution	2-E	.0356	Aesthetic Value of Right-of-Way	2-C	.0346
16	Visual Quality of Highway	2-A	.0350	Housing	5-D	.0345
17	Personal or Group Stress	6-D	.0347	Aesthetic Aspects of Pollution	2-E	.0345
18	Historic Sites	5-E	.0317	Municipal Services	4-A	.0323
19	Barrier Effect	7-D	.0307	Community Cohesiveness	7-A	.0304
20	Municipal Services	4-A	.0305	Visual Quality of Highway	2-A	.0302
21	Financial Capability of Government	4-C	.0295	Barrier Effect	7-D	.0291
22	Neighborhood Stability	7-C	.0267	Satisfaction with Government	4-E	.0286
23	Community Cohesiveness	7-A	.0254	Community Security	4-D	.0255
24	Community Security	4-D	.0251	Community Oriented Contacts	7-E	.0251

Table 6-7 (cont.)

Item Rank	Detrimental	Category	Weight	Beneficial	Category	Weight
25	Public Participation in Government	4-B	.0236	Personal or Business Contacts	7-B	.0235
26	National Defense	6-B	.0230	National Defense	6-B	.0227
27	Personal or Business Contacts	7-B	.0222	Public Participation in Government	4-B	.0217
28	Satisfaction with Government	4-E	.0214	Temporary Economic Effects	3-D	.0212
29	Temporary Economic Effects	3-D	.0184	Historic Sites	3-D	.0206
30	Temporary Aesthetic Effects	2-D	.0117	Temporary Aesthetic Effects	2-D	.0157
Category Rank	**Detrimental**	**Category**	**Weight**	**Beneficial**	**Category**	**Weight**
1	Economics	8-B	.2022	Economics	8-B	.2011
2	Land Use	8-D	.1930	Land Use	8-D	.1865
3	Health and Safety	8-E	.1747	Health and Safety	8-E	.1734
4	Aesthetics	8-A	.1587	Aesthetics	8-A	.1500
5	Social-Psychological	8-F	.1408	Social-Psychological	8-F	.1450
6	Political	8-C	.1302	Political	8-C	.1437
General Rank	**Detrimental**	**Category**	**Weight**	**Beneficial**	**Category**	**Weight**
1	Nonuser Effects	1-B	.6111	User Effects	1-A	.5013
2	User Effects	1-A	.3888	Nonuser Effects	1-B	.4986

the result that the development of new highway networks is often publicly advocated on the basis that the mobility and access provided by the facility will generate economic development benefits in the area of the proposed corridor.

Other factors also attracted attention from the highway officials, although for the most part, the emphasis upon these impacts appeared on an item basis rather than as a general importance attached to all of the impacts found in a particular category. Certain design aspects attracted particular attention especially in terms of their potential harmful effect on the community. Also, on the detrimental side, the respondents indicated concern with specific social-psychological effects of road building. It is interesting to note that the items in the health and safety category attracted far less attention from the highway officials than from the public as a whole. There were, however, two impacts in the group for this which was not true. These were: safety on adjacent highways and the effect upon national defense. Both were awarded a rank that corresponded to the one assigned by the aggregate of regional participants. On the beneficial side, the highway officials again placed added emphasis on several of the variables in the social-psychological category and they were attached also an importance to the beneficial community impact on certain key items found in the political category.

While the answers supplied by the highway officials do appear to vary in some instances from those obtained from the public as a whole, there does exist a strong element of similarity in attitudes. The points of variance, however, will attract the most attention in the following profile. In focusing upon these differences and offering hypothesis regarding their origin, those involved in future highway planning will be made more fully aware of the actual trade-offs applied in corridor route selection.

In highlighting the points of conflict, it must be emphasized that the perspective from which the highway department supplied their answers to the questionnaire was slightly different from that of the participating members of the planning commissions. In particular, the highway officials were asked to react to the hypothetical highway example in terms of the state as a whole without concentrating upon any particular region. Also, they were instructed to review these items acting in the professional capacity as they make daily decisions for the bureau. Several of the respondents noted, and correctly so, that their answers would tend to differ slightly as the hypothetical corridor shifted from one area of the state to another. This point was recognized and accepted. However, it was still hoped that the aggregate of answers supplied by the planning commission member would, as a whole, tend to balance out. This might allow their results to more closely approach the concept of the state used by the highway officials. In assuming the elimination of this distorting effect caused by regional differences, it would then be possible to focus upon the actual points of difference, and to use these as inputs into any future revision of the planning process.

Analysis of the Detrimental Results

On the detrimental side, the highway officials were most concerned with the potential harm that a corridor could do economically to the surrounding community. This is shown by the first, second, and third ratings assigned to the effect upon the level of income, property values and the number of jobs. This concern with economics, while strong throughout each of the regions of the state, was not emphasized to the precise degree that it was by the highway officials. Only one other region felt that income was the most important variable and the sixth, second, and forth ratings assigned to these impacts by the aggregate of regional respondents indicates that they felt there were other items of greater importance when considering the impact of highway construction.

This concern with economics on the part of the highway officials extends to the impact of potential construction upon the number of welfare recipients in the community. While this item rated eighth in the highway hierarchy, it received only an eighteenth rank attention from the state aggregate. This higher level of official concern may well be justified due to two forces. First, the commission members participating in the survey were generally biased toward the property owning middle class, who tend to hold a lower regard for those that are unable to find adequate employment. Consequently, the under representation of minority groups, where unemployment is highest and who are found mostly in the core city, may have placed this particular item lower in the state aggregate than it would actually appear if random sampling techniques were applied. Secondly, the payment of welfare funds is predominately a state rather than a regional or local function. Consequently, officials at the state level may be more accutely aware of the hardships, both personal and financial, that the aggravation of this problem can cause.

The items that on the state level appear equal to, if not more important than, economic considerations are generally found in the health and safety category. There is a significant discrepancy in the ranks assigned to the health aspects of pollution and the access to emergency facilities between the state as a whole and the highway officials. With respect to both items, they attracted more attention from the commission members than from the highway officials. This is especially true of the health aspects of pollution, which rated first in the state and thirteenth among highway officials. This lower regard for the problem of pollution is also evidence to a lesser extent by the ninth rating assigned to the aesthetic effects of pollution by the state and the fifteenth ranking it received from highway planners. With respect to these two items, it appears that there is a lag between the growth in importance of these impacts on the part of individuals and the recognition of this importance by highway officials. Highway officials, however, feel that the lower ranking given to the health effects of pollution is due to their awareness of the problem and solutions relative to other possible impacts. This lag is evidence of the flow nature of community attitudes, which change over time with fluctuations in both surrounding circumstances and the information inputs of individuals. These variations are capable of altering the importance of different items. Apparently the area of pollution is one where highway officials might do well to notice the discrepancy and either make corrections or offer acceptable explanations regarding why greater importance is attached to other potential impacts.

A different hypothesis can be offered to partly account for the discrepancy between the third rating assigned to the access to emergency facilities on the part of the state and the tenth rating awarded to it by the highway planners. The greater importance attached to this item by commission members may be an indication that they view this item as directly applicable to their own situation. Consequently, they feel a personal threat when access is not available in times of crisis. On the other hand, the highway planners may have been regarding this item from a statistical or overall point of view, which allowed them a greater degree of detachment. This detachment places greater emphasis upon the fact that there is only so much access which can be provided. And consequently, a certain amount of risk must be accepted as the result of any corridor choice. To a lesser extent, this state wide concern with items that directly threaten the individual is evidenced by the fourteenth rating associated with the effect a highway has on personal or group stress, as opposed to the seventeenth ranking assigned to it by highway officials. Again, while being sympathetic to human needs, the highway officials were viewing the problem both in the aggregate and from an objective or statistical point of view. Consequently, they might be willing to allow for a lower weight attached to the potential harmful effect of this item.

In the land use category, the pattern of response is similar to that of the economic group, with highway officials directing greater attention in general toward the items than the commission members did. This is evidenced by the fourth, seventh, and sixth rankings assigned to the pattern of land development, the number of business firms and the amount of open space, which received only a seventh, tenth, and eighth ranking state wide. To a lesser extent, greater official concern was also directed toward the effect upon historic sites as evidenced by the eighteenth as opposed to the twentieth rating assigned to it.

As indicated in the introductory statement, this emphasis upon land use mirrors the traditional focus of interest that highway departments have had in the past. The land use variable in the aggregate serves as the focal point of several transport planning models, which try to forecast future road construction needs. Quite possibly, the observed variance in answers serves as an indication that planners should place less emphasis upon land use patterns and redirect their attention more towards the personal or community action aspects of highway construction. This would be especially true as local control over land use patterns through zoning and other measures progresses to the point that highway construction may not have the projected economic effect on the community without approval from local authorities.

With respect to the added concern attached to the number of historic sites, this might have been justified on the part of highway officials because they are more aware of the interests that can be generated within a community if sites possessing historic value are disturbed by highway construction. This point is reinforced by the observation noted in several regions that if people are aware of the historic nature of their area, they become more concerned about actually protecting specific sites.

In the aesthetic category, the highway officials tended to attach a slightly greater importance to the design items of blending the highway into the background and the aesthetic value of the right-of-way as shown by their ninth and twelfth ranking compared to the twelfth and the thirteenth for the state as a whole. This too might be partially attributed to the flow aspect of the value structure which in the past has emphasized the design aspects of highway construction. This has reached the point that highway officials have placed a greater emphasis in current construction on these items than is warranted by public concern. In general, it would appear that the aesthetic concern has shifted from purely design aspects to the external effects of the highway, particularly the impacts of noise, air, and soil pollution upon the community. While design impacts are important, it might appear fruitful for the highway bureau to afford greater attention to other external effects of highway construction.

It is interesting to note, however, that with respect to aesthetic items, the highway bureau is within the range of variation established by the responses from the individual regions. This point is not true, however, with respect to the health and safety factors of access to emergency facilities and the health aspects of pollution. Consequently, it would be safe to estimate that the latter items are the ones deserving greater attention from highway planners.

In the social psychological category, two items were singled out for greater attention on the part of highway officials. These were the barrier effect which ranked nineteenth and community oriented contacts which ranked fourteenth. Both of these relate to the neighborhood aspects of the facility and indicate that highway planners are conscious of the social implications that improperly located routes can have on the affected towns. The positioning of these two items may have been more in line with what a random sample would have shown than the aggregates obtained from the commission members. The middle-class suburban orientation of the participating regional representatives would tend to have them place less emphasis upon the neighborhood or social aspects of their community of residence. With a higher degree of mobility and a wider range of nondaily contacts, they may not feel as threatened by a disruption in these linkages as those located in the urban core areas. Also, the history of harmful effects on these two factors has been in the urban areas, which were under representative in the sample group. As a result, added attention may have been properly directed towards these particular impacts.

Finally, in the political category, highway officials showed even less concern for this group of items, which were generally neglected by the participating regional representatives. This is especially true of the twentieth and twenty-first rankings assigned to the effect the highway might have on municipal services or on the financial capabilities of government. These items, which ranked seventeenth and nineteenth in the state, indicate either that highway officials do not see them as being severely affected by potential route selection or they do not regard them as items that deserve attention within the planning process. Clearly, however, there is a relationship between the potential effect upon property and the ensuing relationship that this property has to the tax base of the community. Furthermore, since it is the tax base that provides the local funding for municipal services, it would appear that the highway bureau as well as the commission members overlooked the essential nature of these two items. Part of the explanation for this may be a function of past highway bureau experience thus leading to the problem of possible versus probable impact occurrence mentioned earlier. With respect to the other impacts found in the political category, the highway officials appear to be compatible in their replies with the answers supplied by the aggregate of commission members.

Analysis of the Beneficial Results

On the beneficial side, the highway officials again placed prime emphasis on the economic considerations by ranking the effect upon the level of income, the effect upon the number of jobs, and the effect on property values first, third and fifth in the hierarchy. This pattern of response is in line with that of the aggregate of commission members as shown by the second, third, and seventh rankings assigned to the same items. Clearly, both groups see some of the major benefits of highway construction being economic spin-offs that favorably affect the development of the region. The only real discrepancy is the greater importance attached to the number of jobs by the state aggregate as opposed to the prime concern allocated to the level of income by the highway officials.

In weighting the variables that are to modify the economic effects of highway construction, the highway planners placed considerable emphasis upon several of the land use items, and in doing this, their rankings differ from those of the state aggregate. This is shown by the second and fourth rankings assigned to the pattern of land development and the effect upon open space, which received a fifth and eighth rating in the aggregate. Clearly, the highway planners placed a greater value upon controlling the physical environment as it relates to the spread of economic development. This in part reflects the traditional pattern of highway planning.

In opposing this view, the regional representatives felt that any economic development should be modified by personal health and safety concerns. This is evidenced by the first, fourth, and sixth rankings assigned to the health effects of pollution, the access to emergency facilities, and safety on adjacent highways. The highway officials showed a lower level of concern for the first two impacts as witnessed by the ninth and tenth rankings assigned to them. With respect to the health aspects of pollution, the variance might again be symptomatic of the fluctuation of values over time with public concern growing regarding the polluting effect of automobile travel on the environment. Regarding the access to emergency facilities the variation might be attributed to the point of view that the respondent used in answering the questionnaire. The regional participant obviously saw himself being affected by these mobility problems at times of crisis. Consequently, where the personal threat is evident, there is a tendency to rate the item higher on the value scale. On the other hand, the highway bureau looks at these items more in terms of statistics; recognizing the physical limitations attached to accessibility, they assigned it a lesser rating. One area of approximate agreement was the beneficial effect on safety on adjacent highways. Here it would appear that the personal concern demonstrated by the regional respondents was equally matched in its intensity by the feelings on the part of highway officials. Planners tend to recognize that this particular item is within their control and consequently, it does receive substantial attention in the decision-making process.

In the aesthetic category, the highway planners and regional representatives attached approximately the same level of concern to the design aspects of highway construction with the major item of variance being blending the highway into the background which attracted more attention from the officials. Another source of controversy was the aesthetic effects of pollution, which was ranked seventeenth by the highway officials and tenth by the state as a whole. The latter again appears to support the lag hypothesis regarding the recognition on the part of highway officials that this item has grown to be a factor of important relative ranking to be considered within the corridor route location decision.

In the political category, the highway officials did single out the financial capabilities of government as being an item deserving attention in the route location decision. This appears related to the emphasis upon the economic and land use effects of highway construction, in that planners feel the principle governmental effect of these impacts is to favorably alter the tax base of the community. Curiously enough, highway planners are cognizant of the interrelationship of these items on the beneficial side, but apparently seem less concerned with their existence if they harm the community. Unfortunately, this is true, although to a lesser degree, of the regional participants as well.

Finally, in the social-psychological category, the highway officials did single out one item for particular attention relative to that allocated by the state aggregate. This is shown by the eleventh position assigned to neighborhood stability as opposed to the twenty-first blank rating it received from the state as a whole. This again may refer, on the part of the highway officials, to the effect of construction within urban communities as opposed to suburban areas where the communications value of neighborhood stability might not be important. Consequently, with the middle-class suburban bias in the sample, there is some indication that the highway officials may have been more correct in awarding this particular item greater attention in their set of trade-offs.

On the other hand, the highway planners did allocate less attention to two community cohesiveness and community oriented contacts nineteenth and twenty-fourth as opposed to fifteenth and twentieth. Quite possibly this may indicate that the greater importance attached to neighborhood stability actually refers to the physical condition and ownership of property rather than the maintenance of social contacts. Consequently, if it does, highway planners might do well to reappraise their thinking regarding the beneficial social-psychological effects of highway construction.

6.4 Analytical Profile of the Weights Assigned by the Regional Planning Directors

This section contains an analysis of the weights obtained from the individual regional planning directors. There were twelve directors participating in this phase of the study, with the sole omission from the list of operating planning regions being accounted for by the director of the valley regional planning agency. These weights reflect the aggregate of answers supplied by the group and represent their collective views regarding the importance of the impacts as they relate to the hypothetical highway project. The answers were initially supplied by the planning director using his own region as a frame of reference. The ratings established through this procedure were then weighted by the percentage of state population found in the region, and these individual weighted results were then aggregated up to provide the planning director's profile of the state as a whole.[1]

In general, the pattern of response indicates a strong concern on the part of the planning directors for the majority of items found in the health and safety category. These items are of dominant concern whether the proposed impact is of a beneficial or a detrimental nature. This set of answers is significant in that the planning directors are generally a detached objective group, who replied to the questionnaire by estimating the importance of a potential impact on the community from the professional point of view. The element of individual threat attached to the health and safety items was not present as it was with the commission members. Consequently, the concern attached to these items cannot be interpreted as a personalized response to potential crisis or emergency. Yet the items still attracted the heaviest and most persistent concern generated by the planning directors.

In modifying their concern with health and safety items, the planners were better able to select specific items from the categories than were either of the two other participating units. The importance that they attached to a collection of items was not overwhelming in any area, although the economic category could be considered as the one receiving the most overall attention from the planning directors. However, even this observation would have to be modified somewhat on the detrimental side as both aesthetic and social-psychological items received relatively strong attention. This is shown by the fact that items from both of these categories appear in the list of the ten most important items. Curiously enough, each category also contains items in the lowest ten, which clearly indicates an ability on the part of the planners to single out the impacts that they felt deserved lesser attention.

Surprisingly, the physical items contained in the land use category received relatively minor weight on the detrimental side, with some improvement appearing as the planning directors considered the impact as being beneficial to the

Table 6-8
Weighted Coefficient and Order of Rank for Each of the Thirty Potential Impacts (Group: Planning Directors)

Item Rank	Detrimental	Category	Weight	Beneficial	Category	Weight
1	Health Pollution	6-C	.0486	Jobs	3-A	.0510
2	Community Stress	6-D	.0444	Health Pollution	6-C	.0503
3	Jobs	3-A	.0418	Community Stress	6-D	.0475
4	Neighborhood Stability	7-C	.0413	Housing	5-D	.0431
5	Blend Highway into Background	2-B	.0407	Safety on Adjacent Highway	6-E	.0429
6	Safety on Adjacent Highway	6-E	.0401	Income	3-E	.0421
7	Aesthetic Aspects of Pollution	2-E	.0395	Accessibility of Emergency Facilities	6-A	.0399
8	Accessibility of Emergency Facilities	6-A	.0391	Community Cohesiveness	7-A	.0396
9	Property Values	3-C	.0391	Property Values	3-C	.0390
10	Community Cohesiveness	7-A	.0373	Welfare Recipients	3-B	.0380
11	Visual Quality of Highway	2-A	.0372	Pattern of Land Development	5-A	.0379
12	Housing	5-D	.0370	Neighborhood Stability	7-C	.0363
13	Barrier Effect	7-D	.0369	Blend Highway into Background	2-B	.0347
14	Income	3-E	.0367	Aesthetic Aspects of Pollution	2-E	.0347
15	Welfare Recipients	3-B	.0355	Open Space	5-C	.0345
16	Financial Capability of Government	4-C	.0337	Community Oriented Contacts	7-E	.0332
17	Open Space	5-C	.0337	Business Firms	5-B	.0333
18	Community Oriented Contacts	7-E	.0325	Financial Capability of Government	4-C	.0305
19	Pattern of Land Development	5-A	.0320	Visual Quality of Highway	2-A	.0300
20	Aesthetic Value of Right-of-Way	2-C	.0317	Aesthetic Value of Right-of-Way	2-C	.0298
21	Personal or Business Contacts	7-B	.0313	Barrier Effect	7-D	.0298
22	Municipal Services	4-A	.0280	Personal or Business Contacts	7-B	.0297
23	Participation in Government	4-B	.0278	Community Security	4-D	.0282
24	Satisfaction with Government	4-E	.0263	Municipal Services	4-A	.0276

Table 6-8 (cont.)

Item Rank	Detrimental	Category	Weight	Beneficial	Category	Weight
25	Business Firms	5-B	.0262	Participation in Government	4-B	.0253
26	Historic Sites	5-E	.0254	Satisfaction with Government	4-E	.0237
27	Community Security	4-D	.0203	Historic Sites	5-E	.0183
28	Temporary Economic Effects	3-D	.0198	Temporary Aesthetic Effects	2-D	.0165
29	National Defense	6-B	.0175	National Defense	6-B	.0160
30	Temporary Aesthetic Effects	2-D	.0168	Temporary Economic Effects	3-D	.0150
Category Rank	**Detrimental**	**Category**	**Weight**	**Beneficial**	**Category**	**Weight**
1	Health and Safety	8-E	.1900	Health and Safety	8-E	.1967
2	Social-Psychological	8-F	.1795	Economics	8-B	.1853
3	Economics	8-B	.1731	Social-Psychological	8-F	.1689
4	Aesthetic	8-A	.1662	Land Use	8-D	.1673
5	Land Use	8-D	.1545	Aesthetic	8-A	.1459
6	Political	8-C	.1364	Political	8-C	.1355
General Rank	**Detrimental**	**Category**	**Weight**	**Beneficial**	**Category**	**Weight**
1	Nonuser Effects	1-B	.4021	Nonuser Effects	2-B	.5163
2	User Effects	1-A	.5978	User Effects	2-A	.4836

community. Also, the governmental items received little attention on the detrimental side, with no apparent improvement in importance on the beneficial side.

It is apparent from their answers that the planning directors are most concerned with the humanistic items, which refer to the quality of life within the state. While this is true on both sides of the ledger, there is a recognition that the economic base of a region is an item to be considered very carefully in highway planning, and one that does possess a special relationship to proposed highway projects. With respect to land use items, however, this recognition of justified concern is not present and this may indicate that the planners feel too much attention has been devoted to this particular set of variables, to the extent that the humanistic or social impacts of highway construction have been neglected in the past.

This humanistic trait is a fairly consistent characteristic and may be a reflection of their general attitude that directors must be in the forefront of planning to educate and guide their commission members and state officials towards the recognition of social deficiencies and the need for a reallocation of resources. In doing this they strive for the perceived imbalances in the present decision-making process.

Clearly, given the opportunity, their pattern of corridor locations would be far different from either their commission members or the state highway officials. Such a deviation would appear most strongly in urban areas, where the anticipated impact upon people would be potentially both the largest and the most severe. Consequently, in assessing and applying the comments of the planning directors as evidence regarding the quality of the answers supplied by the commission member, this particular humanistic trait must be considered. In retrospect, it would appear that the planning directors were able to evaluate the responses of their commission members on an impartial basis. They were cognizant of the variation between their answers and those of the commission members. Consequently, they were careful to disassociate their personal views from the analysis of their members' responses.

Analysis of the Detrimental Results

Clearly, the most important items were found in the health and safety category. This is evidenced by the first and second ranking attached to the health aspects of pollution and personal or group stress. To a lesser extent, this is also reflected in the sixth ranking assigned to the safety on adjacent highways and the eighth ranking assigned to the access to emergency facilities. With respect to two of the items, the directors were in obvious agreement with their planning commissions in assessing the importance of the potential impact to the community as a whole. This is shown by the first and fifth weight assigned by commission member to the health aspects of pollution and the safety on adjacent highways.

The principal point of departure (and its appearance is obviously significant) relates to the rank assigned to the personal or group stress. Only two of the regions felt it was important enough to include it within the top ten, and the fourteenth rating throughout the state indicates it was not an item commanding dominant concern. On the other hand, the planning directors, since they are familiar with current conditions within their regions, may have felt that the present level of tension or stress was already significant. Consequently, any magnification of present problems, especially the urban-suburban conflict, could hold dire consequences for the future stability of the area. Being professionals trained in the development of social relationships and the projection of potential community impacts, they might be well aware of the estimated effect that changes in this item can have if existing problems are not corrected or are actually aggravated.

Regarding the other four items, the extent of agreement between the planning director and the commission members is rather surprising, considering that the directors took the questionnaire before it was administered to the members of the commission, which eliminated the possibility of common answers brought about by external forces. Clearly, the concern of planning directors for individuals within the region is at least as strong as that held by the individuals themselves.

In the economic category, only one impact achieved the high ranking assigned to it by the commission members as a whole. The effect upon the number of jobs was rated third by the planners and fourth in the state as a whole. On the other hand, property values and the level of income, rated ninth and fourteenth by the planners, were considerably more important to the commission members as shown by the second and sixth ratings within the state. With respect to property values, the variance was such that no individual region came close to rating the item as low as the planning directors did. This may be due in part to the property owning bias that was introduced into the sample group. This middle-class suburban group would naturally see the potential for property damage as a threat to their own personal positions. Consequently, the planners viewing the corridor from a broader perspective may have been partially correct in assigning this particular item a lower value in the hierarchy.

The level of income may not have received the proper amount of attention from the planners because they viewed this particular item in the aggregate and, as a result, felt that the loss of income in one town within a region may be balanced by its transfer to another. This broad perspective would not, however, be applicable to the individual respondent. Consequently, since the commission member answered with their own town and community in mind, their answers may have tended to be more personalized reflecting the revenue and development loss experienced by their town of residence.

In the aesthetic category, several items stand out. The first is the seventh rank assigned to the aesthetic effects of pollution, which is in agreement with the ninth rank assigned in the state as a whole. On the other hand, by assigning the fifth rank to the design aspect of blending the highway into the background the planners were showing a far greater concern with this item than was indicated by the twelfth rank assigned by the aggregate of commission members. This greater concern for design is also apparent in the eleventh rank assigned to the visual quality of the highway, which appeared fifteenth within the state. This is an obvious indication on the part of the planning directors that these design items must be considered more heavily than they have in the past. The aesthetic quality of the environment must be protected from destruction by man-made facilities that infringe upon the natural contour of the land.

Curiously, this concern with design aspects of highway impact does not extend to the importance attached to the aesthetic value of the right-of-way which was ranked twentieth by the planners as opposed to thirteenth in the state. This may be a recognition on the part of the planners that, where highway construction has bordered on either natural or man-made regions, the tendency has been to use the right-of-way in proper amounts so that the property taken either improves its use in highway and/or minimizes the extent of loss experienced.

The social-psychological category attracted considerable attention as shown by the fourth, tenth, and thirteenth ratings assigned to neighborhood stability, community cohesiveness, and the barrier effect of highway construction. These three items are of far greater importance to the planners than they are to the commission members. Again, this may be due to the urban nature of two of the items, namely neighborhood stability and the barrier effect, which may not have received the proper amount of attention from a suburban oriented sample group. Clearly, existing social linkages within the community are usually more important in the urban area, where personal mobility relative to that in suburban regions is generally less. Consequently, importance attached to neighborhood stability and the barrier effect, indicate the need for maintaining neighborhood lines of communication.

Even the eighteenth ranking assigned to community oriented contacts is a symptom of the more urbanized point of view held by the planning directors and their recognition of the need to maintain existing social contacts or the effect upon the community that their destruction would have. The community cohesiveness rank, on the other hand, is more a reflection of the regional function of the planning director. He is concerned with people coming together in decision-making groups to offer opinions and ideas that will indicate the direction community action should take. In order to solve problems and achieve goals, this element of cohesiveness must be there and its destruction or damage as a result of highway construction would, in the planner's opinion, be a severe blow to the orderly development of the region.

The land use category spotlights an area of wide divergence between the responses of the planners and the replies of the commission members in the aggregate. In particular, the nineteenth, twenty-fifth, and seventeenth ranks assigned to the pattern of land development, the number of business firms, and the amount of open space indicate that planners are willing to sacrifice the physical effect of the highway on an area, where the gain is to be measured in terms of protecting the social structure from external harm. Moreover, in addition to this relative evaluation, it may be that the planners are indicating that more attention has been devoted to land use variables in the past than is warranted by their relative impact upon the community. Consequently, present emphasis should shift away from land use effects and to the more personal impacts of construction, which have tended to be neglected in the past.

The only real point of agreement between the two groups was in the area of the potential impact upon housing units, which was ranked twelfth by the planners and eleventh by the commission members. Obviously, these two groups, along with the highway officials, are fairly unanimous in their assessment of the importance of housing compared to other impacts.

Finally, in the political category, the general attitude of the planners was to play down the importance of these items except where the financial capability of government was concerned. This item placed sixteenth, which allowed it to appear the closest to the state ranking. With respect to the other items, the regional planners were less concerned with the effect on municipal services and more concerned that an improperly constructed highway might have a harmful impact on the level of satisfaction that individuals have with government actions. This latter difference, although the item itself is still minor in importance, may be an indication of the planner's recognition that difficulty with past actions makes future plans harder to implement, since the public builds up a wall of resistance to government-oriented forces of change.

With respect to both the financial capability of government and municipal services, it may be that the planners answering the question on a regional basis did not see significant harm accruing from highway construction in the area. Regarding the financial capabilities of government, many planners spoke of industry being shifted around within the region, which would not affect the overall tax base but which might, however, have a significant effect on individual towns. As a result, municipal services would not be severely curtailed on a regional basis, either through disruption or through a lack of funding.

Analysis of the Beneficial Results

On the beneficial side, the same pattern of response was observed regarding the items of overall importance contained in the health and safety category. However, there was a marked change in the particular modifiers that were selected from the supporting groups. The health and safety group again places four items in the top ten, with the health aspects of pollution, personal or group stress, safety on adjacent highways, and the access to emergency facilities ranking second, third, fifth, and seventh in the hierarchy. As on the detrimental side, all these rankings are fairly consistent with those assigned by the regional commission members, except the importance attached to personal or group stress. Obviously, the regional planning directors see a far greater threat from this particular problem, either because of its existing magnitude or its potential for future community disruption. Consequently, highway construction, if it can reduce community or personal problems, should be rated very strongly in the decision-making process.

Three items contained in the economic category gain considerable importance relative to the positions assigned to them on the detrimental side. This is evidenced by the first, sixth and tenth rankings assigned to the effect upon the number of jobs, the effect upon the level of income, and the effect upon the number of welfare recipients. With the rankings assigned to these items and the importance attached to property values, the regional planners were in partial agreement with the members of their commission. This is especially true regarding the prime weight attached to the number of jobs. The tenth rank assigned to the number of welfare recipients may be further evidence of the humanistic attitude of the planning directors. This point can be combined with the fact that the ranking might be a correction factor deliberately introduced to counteract the suburban orientation of the participating commission members. It is interesting to note that even with this compensating factor, the rank assigned to the item still is below that assigned by state highway planners. It is possible that with respect to this item, the highway officials have a greater sensitivity for the effect it can have on the community as a whole.

In the land use category, most of the items receive less attention from the regional planners than they do from the members of the planning commissions. This is especially true of the eleventh rank assigned to the pattern of land development, the fifteenth rank assigned to the amount of open space, and the seventeenth rank assigned to the number of business firms. Each of these three items rated far more important within the individual regions and, as a result, for the state as a whole.

A variation in the opposite direction is observed with respect to the importance attached to the beneficial effect on the number of housing units, which was ranked fourth by the regional planners and eleventh within the state. Here, the planners' recognition of the extent of this problem not only regionally, but throughout the entire state, may have influenced their weighting of the item. Moreover, the commission members, for the most part, were already homeowners who have not recently experienced the tightness of the housing market. Consequently, their awareness of this problem is indirect rather than direct. As a result, the importance they attach to this item may be less than it actually deserves considering the present condition of the housing market in the state.

In the social-psychological category the community oriented items again receive a far stronger rating from the regional planning directors than they do from the commission members. Community cohesiveness places eighth, while neighborhood stability and community oriented contacts are twelfth and sixteenth in the hierarchy of the regional planners. Again, this shows a concern with the social environment that has generally been neglected by highway planners. Too often, past highway construction has neglected the social good that can be generated within the community by proper route location planning. Consequently, this particular area has been downgraded in favor of land use and economic considerations. However, several planners in their verbal comments noted that properly constructed highways increase both personal and social linkages by improving access and mobility. These result in an improved regional feeling that is better equipped to deal with the problems which extend beyond individual towns. By placing emphasis upon this communications aspect of highway construction, the planners are indicating that, where possible, this particular item should be given considerable weight within the planning process.

In the aesthetic category, the regional planners showed a mixed level of concern with the design aspects of highway construction relative to that indicated by the commission members. The groups attached the same relative rank to the visual quality of the highway, while the planners were slightly more concerned with blending the highway into the background and somewhat less interested in the aesthetic value of the land used for the right-of-way. With respect to the latter, it may be that commission members saw the use of undesirable land for right-of-way purposes, not only as an aesthetic improvement, but also in terms of limiting the expenditure of local tax revenue, where the land taken is of poor or blighted urban buildings.

One item of surprising interest was the fourteenth rank assigned to the aesthetic effects of pollution by the planners, as opposed to the tenth rating it received in the state as a whole. This would indicate that, while planners are concerned with health aspects which rated second, they realize there are other more important factors. Moreover, this may be an indication that the planners are more aware of the fact that in order to experience development and growth in the region, the residents of the area must be willing to accept a lower level of aesthetic improvement.

Finally, in the political category the planners were even less concerned than the regional respondents with the connection between highway construction and the government's relationship with the public. The two key items of the financial capabilities of government and municipal services rated eighteenth and twenty-fourth, showing a general lack of concern. Again, this may be due to the fact that planners were replying on a regional rather than a town basis. Consequently, they might be aware that for the region as a whole, fewer administrative benefits might be derived from highway construction.

A plausible alternative hypothesis rests upon the relatively low value that planners attached to attracting new business firms as well as the lesser relative importance awarded to property values. This hypothesis is that these items in the past have received an overabundance of attention from highway officials and that by de-emphasizing these items the planners are indicating that there are impacts upon the social and personal environment that deserve far greater attention in route location process. In other words, by simply expanding the level of physical services provided to community residents, the decision-making process should not be justified in overlooking the effect that a highway can have upon social or personal linkages, especially in urban areas.

Notes

Notes

Chapter 1
Survey of the Literature

1. Nicholas Kaldor, "Welfare Propositions in Economics and Interpersonal Comparisons of Utility," *Economic Journal* 49 (September 1939): 549-52; J.R. Hicks, "The Foundations of Welfare Economics," *Economic Journal* 49 (September 1939): 696-712.

2. For a general survey of cost-benefit analysis see: A.R. Prest and R. Turvey, "Cost-Benefit Analysis: A Survey," *Economic Journal* 75 (December 1965): 683-735; P.D. Henderson, "Notes on Public Investment Criteria in the United Kingdom," *Bulletin of the Oxford University Institute of Economic and Statistics* 27 (February 1965): 55-89. Reprinted with amendments in *Public Enterprise*, ed. by R. Turvey (Baltimore, Maryland: Penguin Books, 1968), 86-169; Joseph D. Crumlish, "Notes on the State-of-the-Art of Benefit-Cost Analysis as Related to Transportation Systems" (United States Department of Commerce, National Bureau of Standards, GOP: 1966).

3. Prest and Turvey, "Cost-Benefit Analysis," p. 690.

4. Oregon State Highway Department, "The Economics of Highway Planning," *Technical Bulletin No. 7* (Salem, Oregon: 1937).

5. See Robert Hennes, *Highway Research Board Special Report 56* (Highway Research Board, 1960), pp. 131-35, for the history of the need for such a scientific device.

6. American Association of State Highway Officials, *Road User Benefit Analysis for Highway Improvements, Part II.* (Committee on Planning and Design Policies, AASHO, Washington, D.C., 1960).

7. Ibid., p. 10.

8. Bradford Sears, "Highways as Environmental Elements," *Highway Research Record* No. 93 (Highway Research Board, 1965), p. 49.

9. AASHO, *Road User Benefit*, p. 10.

10. Ibid., p. 11.

11. Ibid.

12. E.L. Grant and C.H. Olisby, "A Critique of Some Recent Economic Studies Comparing Alternative Highway Locations," *Highway Research Board Proceedings* 39 (1960): 1-8.

13. A brief example may aid in demonstrating this point. Assume that the net social value of a particular proposal is determined by comparing the ratio of total benefit to total cost (i.e. B/C). Assume, furthermore, that before social costs are added to the accounting framework that one of the relevant alternatives (X) possessed the following dollar costs and benefits:

	B	C	B/C
X	$20	$10	2

If social costs amounting to $2 are included as positive values on the cost side, then the B/C alters:

X	$20	$12	1.67

If, on the other hand, social costs are recorded as negative benefits, the resulting ratio is appreciably altered:

X	$18	$10	1.8

Inadequate considerations of this possibly might seriously affect the relative ranking of the alternatives.

14. T.M. Coburn, M.E. Beesley and D.J. Reynolds, "The London-Birmingham Motorway: Traffic and Economics," *Road Research Laboratory Technical Paper No. 46* (D.S.I.R., H.M.S.O.: 1960).

15. Prest and Turvey, "Cost-Benefit Analysis," p. 703.

16. Henderson, "Public Investment Criteria," pp. 97-118.

17. R.N. McKean, *Efficiency in Government Through Systems Analysis* (New York: John Wiley & Sons, 1958).

18. J. Hirshleifer, J.C. deHaven and J.W. Milliman, *Water Supply, Economics, Technology and Policy* (Chicago: University of Chicago Press, 1960).

19. AASHO, *Road User Benefit*, p. 126.

20. Robinson Newcomb, "A New Approach to Cost-Benefit Analysis," *Highway Research Record No. 138* (Highway Research Board, 1966), 18-21.

21. A. Myrick Freeman, "Income Distribution and Planning for Public Investment," *American Economic Review* 56, no. 3 (June, 1967): 495-508.

22. Harvey S. Perloff, *How a Region Grows: Supplementary Paper No. 17* (New York: Committee for Economic Development: 1963); idem, "Systems of Economic Accounts and Analysis for Urban Regions: A National System of Metropolitan Information and Analysis," *American Economic Review*, LII No. 2 (May, 1962), 356-64; idem, "Relative Regional Economic Growth: An Approach to Regional Accounts," in *Design of Regional Accounts* (Baltimore, Maryland: Johns Hopkins Press, 1961).

23. Arthur M. Weimer and Homer Hoyt, "Economic Base Analysis" in *Techniques of Urban Economic Analysis*, ed. by R.W. Pfouts (W. Trenton, New Jersey: Chandler Davis Publ. Co., 1960), 20-38.

24. Richard B. Andrews, "Mechanics of the Urban Economic Base," ibid., pp. 40-52.

25. James Giles and William Grigsby, "Classification Errors in Base-Ratio Analysis," ibid., pp. 214-28.

26. Hans Blumenfeld, "The Economic Base of the Metropolis," ibid., pp. 230-77.

27. Charles Levin, *Theorv and Method of Income and Product Accounts for Metropolitan Areas* (Pittsburgh: University of Pittsburgh Press, 1963).

28. Blumenfeld, "Economic Base " in *Techniques of Urban Economic Analysis*.

29. Charles M. Tiebout, "The Urban Economic Base Reconsidered," in *Techniques of Urban Economic Analysis*, 280-324.

30. Hugo Hegeland, *Multiplier Theory* (Lund: C.W.K. Gleerup, 1954).

31. Werner Z. Hirsch, "Design and Use of Regional Accounts" *American Economic Review* 52, no. 2 (May 1962): 365-73.

32. Walter Isard and Robert Kavesh, "Economic Structural Interrelations of Metropolitan Regions," in *Techniques of Urban Economic Analysis*, 360-77.

33. W.L. Hansen and C.M. Tiebout, "An Intersectional Flows Analysis of the California Economy," *Review of Economics and Statistics* 45 (November 1963), 409-18.

34. Charles M. Tiebout, "Regional and Interregional Input-Output Models: An Appraisal," *Southern Economic Journal* 24 (October 1957): 140-47.

35. G. Warren Heenan, "The Economic Effect of Rapid Transit on Real Estate Development," *The Appraisal Journal* 36 (April 1968): 213-24; idem, "The Impact of Rapid Transit on Business and Real Estate in the Central City," (Speech to Market Street Development Project, San Francisco: March 30, 1967).

36. Jay S. Golden, *Land Values in Chicago, Before and After Expressway Construction*, Chicago Area Transportation Study, No. 316, OSI-VI (October 1968), 65 pp.

37. Jere J. Hinkle and Frederick F. Frye, *The Influence of an Expressway on Travel Parameters–The Dan Ryan Study*, Chicago Area Transportation Study, No. 10, 320-VI (June 1965), 49 pp.

38. Walter D. Stoll, *Changes in Average Trip Length: A Case Study by Mode and Purpose of Skokie Trips made in 1956 and 1964*, Chicago Area Transportation Study, No. 311, 014-VI (December 1967), 30 pp.

39. Ibid., p. 26.

40. William H. Dodge, *Influence of a Major Highway Improvement upon the Economy of Dunn and St. Croix Counties, Part I, The Local Economy Before Major Highway Improvements* (Wisconsin Commerce Papers, U. of Wisc.: 1962), 124 pp.

41. Edmond L. Kanwit and Alma F. Eckartt, "Transportation Implications of Employment Trends in Central Cities and Suburbs," *Highway Research Record*, No. 187 (Highway Research Board, 1967), 1-14.

42. David K. Witheford, "Highway Impacts on Downtown and Suburban Shopping," *Highway Research Record*, No. 187 (Highway Research Board, 1967), 15-20.

43. Ibid., p. 16.

44. Paul Diesing, "Non-economic Decision-Making," *Ethics* 66 (October 1955): 18-35; idem, "Socioeconomic Decisions," *Ethics* 69 (October 1958): 1-18.

45. For a description of the role of goals in the planning process, see: Robert C. Young, "Goals and Goal Setting," *Journal of the American Institute of Planners* 32 (March 1966): 70-85.

46. Diesing, "Non-economic Decision-Making," pp. 18-35.

47. A.C. Rogers, "The Urban Freeway: An Experiment in Team Design and Decision-Making," *Highway Research Record*, No. 220 (Highway Research Board, 1968), 20-28; Andrew Euston, "Design Concepts for the Future," *Highway Research Record*, No. 220 (Highway Research Board, 1968), 5-10.

48. Stuart L. Hill, "The Effects of Freeways on Neighboring–Right of Way Research and Development" (Division of Highways, Department of Public Works, Transportation Agency, State of California). In cooperation with the U.S. Department of Transportation, Federal Highway Administration, Bureau of Public Roads (June, 1967); idem, and Banford Frankland, "Mobility as a Measure of Neighborhood," *Highway Research Record* No. 187 (Highway Research Board, 1967), 33-42.

49. Hill, "The Effect of Freeways on Neighborhood," p. 14.

50. Ibid., pp. 16-17.

51. Ibid., p. 23.

52. Raymond Ellis, "Toward Measurement of the Community Consequences of Urban Freeways," *Highway Research Rcord*, No. 229 (Highway Research Board, 1968), 38-52.

53. Ibid., p. 41.

54. Ibid., p. 42.

55. Ibid., p. 43.

56. David S. Gendell, *Evaluation of Alternative Transportation Systems*, Lecture 92 (Bureau of Public Roads), p. 4.

57. Charles C. Schimpeler and William L. Grecco, "The Community Systems Evaluation: An Approach Based on Community Structure and Values." *Highway Research Record* No. 238 (Highway Research Board, 1968), 131-32.

58. Ibid., p. 130.

59. Morris Hill, "A Goals-Achievement Matrix for Evaluating Alternative Plans," *Journal of the American Institute of Planners* 34 (January 1968): p. 24.

60. Ibid., p. 27.

61. Gendell, *Alternative Transportation Systems*, pp. 6-7.

62. Ibid., pp. 6-8.

63. Schimpeler and Grecco, "Community Systems Evaluation," pp. 131-35.

64. William Jessiman, Daniel Brand, Alfred Tumminia, Roger Brussee, "A Rational Decision-Making Technique for Transportation Planning," *Highway Research Record* No. 180 (Highway Research Board, 1967), 71-80.

65. Gendell, *Alternative Transportation Systems*, pp. 9-15.

66. Morris Hill, "A Goals-Achievement Matrix for Evaluating Alternative Plans," *Journal of the American Institute of Planners* 34 (January 1968): p. 19. The discussion of the goal's achievement matrix is based on this article and also Hill's "A Method for Evaluation of Transportation Plans," *Highway Research Record* No. 180 (Highway Research Board, 1967), 21-34.

67. Hill, "Evaluation of Transportation Plans," p. 32.

68. Lyle C. Fitch, comments on "Plan Evaluation Methodologies" by W.A. Steger and T.R. Lakshmann, *Urban Development Models*, Special Report 97 (Highway Research Board, 1968), p. 74.

69. H. Claude Shostal, "Pricing and Planning in Urban Transportation," *Reviews in Urban Economics* (Fall 1968), 5-28.

70. Ibid., pp. 13-14. Also, William Vickery, "Pricing in Urban and Suburban Transport," *American Economic Review* 53, no. 2 (May 1963): 452-65.

71. Wilbur Thompson, *A Preface to Urban Economics* (Baltimore, Md.: The Johns Hopkins Press, 1965).

72. Shostal, "Urban Transportation," *Reviews in Urban Economics* (Fall 1968), 6-7; J.R. Meyer, J.F. Kain and M. Wohl, *The Urban Transportation Problem* (Cambridge: Harvard University Press, 1965).

73. Wilbur A. Steger and T.R. Lakshmanan, "Plan Evaluation Methodologies: Some Aspects of Decision Requirements and Analytical Response," *Urban Development Models*, Special Report 97 (Highway Research Board, 1968), 33-76.

74. Ibid., p. 68.

75. Ibid., p. 71.

76. Charles L. Levin, "Establishing Goals for Regional Economic Development," *Journal of the American Institute of Planners* 30 (May 1964): 100-110.

77. Henry Fagin, "The Penn-Jersey Transportation Study," *Journal of the American Institute of Planners* 29 (February 1963), 9-18.

78. RAND Corporation, "Transportation for Future Urban Communities: A Study Prospectus," RM-2824-FF (Santa Monica, Calif., 1961).

79. J.F. Kain and J.R. Meyer, *A First Approximation to a RAND Model for Study of Urban Transportation*, RM-2878-FF (Santa Monica, California: RAND Corporation, November, 1961).

80. John F. Kain, *A Report on an Urban Transportation Model, Some Progress and Some Problems*, No. B-2549 (Santa Monica, California: RAND Corp., June, 1962), p. 4.

81. Ibid., p. 6.

82. John F. Kain, "The Journey to Work as a Determinant of Residential Location," *Papers and Proceedings of the Regional Science Association* 9 (1962): 137-60.

83. Nancy J. Edin, *Residential Location and Mode of Transportation to Work: A Model of Choice*, Chicago Area Transportation Study No. 311, 012-VI (October 1966).

84. Leon N. Moses, "Towards a Theory of Intra-Urban Wage Differentials and their Influence on Travel Patterns," *Papers and Proceedings of the Regional Science Association* 9 (1962): 53-64.

85. Ira S. Lowry, *A Model of Metropolis*, RM-4035-RC (Santa Monica, California: RAND Corporation, August, 1964).

86. Ira S. Lowry, "Seven Models of Urban Development: A Structural Comparison," in *Urban Development Models*, Special Report 97 (Highway Research Board, 1968), 121-45.

87. Ibid., p. 9-10.

88. Ibid., p. 11.

89. H.W. Bruck, S.H. Putman, and W.A. Steger, "Evaluations of Alternative Transportation Proposals: The Northeast Corridor," *Journal of the American Institute of Planners* 32 (November 1966): 322-33.

90. Duncan MacRae, "An Economic Evaluation of Urban Development." In J.J. Sullivan, editor, Explorations in Urban Land Economics, University of Hartford, 1970, pp. 27-50.

91. Richard E. Quant and William J. Baumol, "The Demand for Abstract Transport Modes," *Journal of Regional Science*, no. 6 (Winter 1966): 13-23.

92. Reuben Gronau and R.E. Alcaly, "The Demand for Abstract Transport Modes: Some Misgivings," *Journal of Regional Science* 9 (April 1969), 153-57.

93. David Braybook and Charles E. Lindbloom, *A Strategy of Decision: Policy Evaluation as a Social Process* (New York: The Free Press of Glencoe, 1963).

94. For a review of the use of systems analysis in the area of defense planning see: C.J. Hitch and R.H. McKean, *The Economics of Defense in the Nuclear Age* (Cambridge: Harvard University Press, 1961).

95. For a general review of the systems approach to evaluation see: *Cost Effectiveness–Economic Evaluation of Engineered Systems*, ed. by J. Morley English (New York: John Wiley & Sons, 1968).

96. See Robert C. Young, "Goals and Goal-Setting," *Journal of the American Institute of Planners* 32 (March 1966): 76-85.

97. For a detailed review of the systems approach see: A.D. Kazanowski, "A Standardized Approach to Cost-Effectiveness Evaluations" in *Cost-Effectiveness*, ed. by J. Morley English (New York: John Wiley & Sons, 1968).

98. Wilfred Owen, *Strategy for Mobility* (Washington, D.C.: Brookings Institution, 1964); idem, "Transportation and Technology," *American Economic Review* 52 (May 1962), 405-13.

99. For a review of the systems approach used in urban transport planning, see Edwin N. Thomas and Joseph L. Schofer, *Strategies for the Evaluation of Alternative Transportation Plans*, National Cooperative Highway Research Program 20-8 (Highway Research Board, 1967).

100. For a detailed discussion of the limitations and fallacies of systems analysis in general see A.D. Kazanowski, "Cost Effectiveness Methodology and

Limitations," NAA/S and ID, SID 64-2829 (1964); idem, "Cost-Effectiveness Fallacies and Misconceptions Revisited" in *Cost Effectiveness*, ed. by J. Morley English (New York: John Wiley and Sons, 1968).

101. For a survey of the general application of systems analysis in the area of public finance see "Program Budgeting and Benefit-Cost Analysis," ed. by Harley H. Hinrichs and Graeme M. Taylor (Pacific Palisades, California: Goodyear Publishing Co., 1969).

Chapter 2
Background for the Model

1. For an example of this type of study see Walter C. McKain, *The Connecticut Turnpike: A Ribbon of Hope*, Storrs, Conn., Storrs, Agricultural Experiment Station, 1964.

2. U.S. Dept. of Transportation, Federal Highway Administration and Bureau of Public Roads, *Policy and Procedure Memorandum* 20-8 (Washington, D.C.: Government Printing Office, 1969).

3. Charles C. Schimpler and William L. Grecco, "The Community Systems Evaluation," *Highway Research Record*, No. 238, pp. 123-52.

4. For an analysis of one of these problems see Helen Leavitt, *Superhighway, Superhoax* (New York: Doubleday and Co.), 1971, ch. 3.

5. For a discussion of this problem as well as other problems found in additional levels of the property rights issue, see Anthony Downs, *Urban Problems and Prospects* (Chicago: Markham Publishing Co., 1970), ch. 8.

6. The concept of amenity rights and their relationship to society as a whole is the subject reviewed by E.J. Mishan, *Technology and Growth: The Price We Pay* (New York: Praeger Publishing Co., 1969).

7. Mishan, *Technology and Growth*, ch. 5.

Chapter 3
The Evaluative Framework: Form and Operation

1. David Braybrook and Charles E. Lindblom, *A Strategy of Decision: Policy Evaluation as a Social Process* (New York: Glencoe, 1963).

2. R.G. Lipsey and Kevin Lancaster, "The General Theory of Second Best," *Review of Economic Studies* 24 (1956-57): 11-32.

3. Lichfield, Nathaniel, "Cost-Benefit Analysis in City Planning," *Journal of the American Institute of Planners* 26 (November 1960): 273-79.

4. Hill, Morris, "A Goals-Achievement Matrix for Evaluating Alternative Plans," *Journal of the American Institute of Planners* 34 (January 1968): 19-29.

5. For a discussion and partial listing of these items see, Anthony Downs,

Uncompensated Non-Construction Costs which Urban Highways and Urban Renewal Impose Upon Residential Households—A paper presented at the Conference on Economics and Public Output, Universities–National Bureau Committee for Economic Research, Princeton University, April 26, 1968.

6. See Ann Fetter Friedlaender, *The Interstate Highway System: A Study In Public Investment* (Amsterdam, Holland: North-Holland Publishing Co., 1965), p. 64.

7. *National Survey of Transportation Attitudes and Behavior: Phase I Summary Report*. National Cooperative Highway Research Program Report 49 (Washington, D.C.: Highway Research Board, 1968); *National Survey of Transportation Attitudes and Behavior: Phase II Final Report*, Project No. HR 20-4 by Robert K. McMillan and Henry Assael, Chilton Research Services–For the National Cooperative Highway Research Program (Philadelphia, Pennsylvania: Highway Research Board, 1968).

8. The Eastern Region includes New England, Middle Atlantic States, Delaware, Maryland, District of Columbia–Phase I Summary, p. 22.

9. Phase II Summary, p. 7.

10. Phase II Summary, pp. 76 (table 11), 83 (table 20), 93 (table 29), 95 (table 38), 100 (table 47).

11. Phase II Summary, pp. 77 (table 12), 84 (table 21), 91 (table 30), 96 (table 39), 101 (table 48).

12. Phase II Summary, p. 7.

13. Phase II Summary, pp. 7-8.

14. Phase II Summary, p. 193.

15. Phase I Summary, p. 19 (tables 38, 39).

16. M.T. Shaffer, "Attitudes, Community Values and Highway Planning," *Highway Research Record* No. 187 (Highway Research Board, 1967), 55-61.

17. F.T. Paine, A.N. Nash, S.J. Hille, and G.A. Brunner, *Consumer Conceived Attributes of Transportation: An Attitude Study*, prepared for U.S. Department of Commerce, Bureau of Public Roads, by Maryland Department of Business Administration (June, 1968).

18. Ibid., p. xi.

19. Paul N. Yevisaker, "The Resident Looks at Community Values," Conference on Transportation and Community Affairs, 15 pp.

20. Gordon Fellman and Roger Rosenblatt, "The Social Costs of an Urban Highway: Cambridge and the Inner-Belt Road," Transportation and Poverty Conference, June 7, 1968, American Academy of Arts and Sciences, 33 pp.

21. Ibid., pp. 26-27.

22. H.K. Dansereau, "Highway Development: Attitudes and Economic Climate," *Highway Research Record* No. 187 (Highway Research Board, 1967), 21-32.

23. Lisa R. Peattie, "Reflections on Advocacy Planning," *Journal of the American Institute of Planners* 34 (March 1968): 80-88.

24. Alan M. Voorhees & Associates, Inc., *Attitudes of Connecticut People* (Hartford: Connecticut Interregional Planning Program, 1966); Charles F. Barnes, "Living Patterns and Attitude Survey," *Highway Research Record* No. 187 (Highway Research Board, 1967), 43-54.

25. Thomas A. Morehouse, "The 1962 Highway Act: A Study in Artful Interpretation," *Journal of the American Institute of Planners* 35, no. 3 (May 1969): 160-68.

26. Policy and Procedure Memorandum 20-8, U.S. Department of Transportation, Federal Highway Administration, Bureau of Public Roads (January 14, 1969), p. 2.

Chapter 4
The Development of a Set of Weights

1. Washington State University, Highway Research Section, Engineering Research Division, *A Study of the Social, Economic and Environmental Impact of Highway Transportation Facilities on Urban Communities* (Pullman, Washington: 1968); Hill, Morris, "A Method for the Evaluation of Transportation Plans," *Highway Research Record* No. 180 (Washington: Highway Research Board, 1967), 21-34.

2. C.E. Ferguson, *Macroeconomic Theory of Workable Competition* (Durham, North Carolina: Duke University Press, 1964).

3. Richard Starr, *Urban Choices* (Baltimore, Md.: Penguin Books, 1967).

4. Norman C. Dulkey, *The Delphi Method: An Experimental Study of Group Opinion* (Santa Monica, California: The Rand Corporation, June, 1969), Memorandum RM-5888-TR.

5. For further information on weighting, please consult the following: Eckenrode, Robert T., "Weighting Multiple Criteria," *Management Science* 12 (November 1965): 180-92; Fishburn, Peter C., "A Note on Recent Developments in Additive Utility Theories for Multi-Factor Situations," *Operations Research* 14 (November-December 1966): 1143-1148; idem, "Methods of Estimating Additive Utilities," *Management Science* 13 (March 1967): 435-53.

Chapter 6
Presentation and Analysis of the Impact Factor Weights

1. The population projections were supplied by the Connecticut Bureau of Highways and were based upon Connecticut Interregional Planning Program Population Projections, Hartford, Connecticut, published in December, 1968.

Bibliography: A Partially Annotated List for Part I

Bibliography

Alonso, William. "Location Theory." In *Regional Development and Planning: A Reader*, edited by J. Friedman and W. Alonso. Cambridge: MIT Press, 1964.

This article is primarily concerned with the location decision from the point of view of the firm, rather than from the point of view of a project planner. Alonso includes a review of the literature regarding location theory, pointing out limitations and deficiencies. He explains the principle of median location and competition along a line for a location decision of a firm with one market and one raw material. He expands the analysis, commenting on how closely the theory resembles reality.

American Association of State Highway Officials. *Road User Benefit Analysis for Highway Improvements: Part I.* "Passenger Cars in Rural Areas," Washington, D.C., 1960.

A manual outlining the procedure to be followed in a benefit-cost study aimed at evaluating the potential net impact of a highway. It focuses on a comparison of user benefits and costs.

Appleyard, Donald and Lynch, Kevin. "Aesthetic Criteria for Transportation Plan Evaluation." In *Strategies for the Evaluation of Alternative Transportation Plans*. Appendix F. National Cooperative Highway Research Program, Project 8-4, July 1967. Edited by Edwin N. Thomas and Joseph L. Schofer.

They evaluate the aesthetic effect of a highway on both the user and the bystander. They offer a set of design criteria to improve the aesthetic quality of the highway for both groups.

Archibald, G.C. "Regional Multiplier Effects in the U.K." *Oxford Economic Papers*, 19 (March 1967): 22-45.

The author discusses a range of possible values for regional multipliers, as the results of his attempt to estimate a minimum value below which the multiplier for any one region could not lie. He analyzes the arguments for migration vs. moving the jobs to help unemployment problems. Setting up a local expenditure model, he finds multiplier effects of increased income and he discusses alternative models in trying to find the effects of migration on the income in the area.

Barkin, David. "Measuring the Regional Impact of Government Spending." Economic Development Administration, Working Paper EDA 10, May, 1968.

He asks if funds are distributed to those regions of the country which are neediest (where the ratio of the number of eligible recipients to need is the highest) concern with micro impact. In general, funds from particular acts are designed to help specific problems, not purely to redistribute income from richer states to poorer states. Thus, it is difficult to compare programs due to differences in composition of people eligible for assistance. He points out the distinction between the effect of government expenditure programs on the spatial distribution of income for the whole population and on the distribution of income between "needy" groups in the population.

Barnes, Charles F. "Living Patterns and Attitude Survey." Highway Research Board, National Academy of Sciences, Washington, D.C., Highway Research Record No. 187, 1967, 43-54.

This study presents the results of four surveys of attitudes concerning town and state features most liked and most disliked, and major problems for the state and town lived in. Also, the survey covers attitudes toward recreation and leisure time. This article is interesting because the data are for Connecticut.

Baumol, William J. "On the Social Rate of Discount." *American Economic Review* 58 (September 1968): 788-802.

He feels that a choice of a single social rate of discount is indeterminant since the corporate tax structure requires a private rate of return nearly twice as large as public investment.

Bellomo, Salvatore and Provost, Steven C. "Two Procedures to Improve the Economic Evaluation of Alternative Highway Systems." Highway Research Board, National Academy of Sciences, Washington, D.C., Highway Research Record No. 224, 1968, 44-53.

Both peak and off-peak traffic loads must be considered if the cost-benefit ratios of alternative systems are to be compared since this will have an effect upon the user cost calculations. Sensitivity analysis should also be applied where small variations in key estimated factors may have a significant effect upon the final choice.

Bevis, Howard W. "The Application of Benefit-Cost Ratios to an Expressway System." *Highway Research Board Proceedings*, 35, Washington, D.C., 1956.

An outline of methodology and items to be considered when comparing alternative systems using a cost-benefit ratio based on annual user benefits and discounted annual costs. He points out that this method is not useful for indicating either fiscal feasibility or for setting the priorities for construction of different segments.

Black, Therel and Black, Jerrilyn. *Some Sociological Considerations in Highway Development*. Highway Research Board, National Academy of Sciences, Washington, D.C., Bulletin 169, 1957, 51-59.

They outline the human factors that must be considered in the construction of a highway. Planning teams should be used to handle intangibles in a nonmonetary way. Minimization of economic costs should not be the sole criterion.

Bloomberg, Warner, and Schmandt, Henry J. "The Issue is Great and Very Much in Doubt." In *Power, Poverty, and Urban Policy*, edited by W. Bloomberg and H. Schmandt. Beverly Hills: Sage Publications, 1968.

With regard to urban poverty, they conclude that we have not reduced poverty to anywhere near the degree that our resources would permit.

Braybrook, David and Lindbloom, Charles E. *A Strategy of Decision: Policy Evaluation as a Social Process*. New York: The Free Press of Glencoe, 1963.

Brenan, Malcom F. and Rothrock, Claude A. "Principles of Highway Engineering Economic Analysis." Highway Research Board, National Academy of Sciences, Washington, D.C., Highway Research Record No. 100, 1965, p. 1.

Describes a manual written by the authors which outlines the alternative and procedure to be followed in calculating the user costs and benefits of highway construction.

Bruck, H.W., Putman, Stephan W. and Steger, Wilbur A. "Evaluation of Alternative Transportation Proposals: The Northeast Corridor." *Journal of the American Institute of Planners* 32 (November 1966): 322-33.

Outlines the use of impact models based on regional input-output tables to derive transportation needs. After the alternatives are listed an evaluation process is needed to select the one which optimizes the desired goals. They list a set of criteria the method must fulfill and outline a group of six possible methods of evaluation.

Buchanan, James M. "Politics, Policy, and the Pigovian Margins." *Economica*, N.S., 29 (February 1962): 17-28.

If private decision leads to effects external to the decision maker, optimum output is not attained through competition even if the remaining efficiency conditions hold. Buchanan questions the assumption that the externalities will fall if the activity is shifted from a market to a political organization. He tries to show that with consistent assumptions about human behavior in market and political institutions, any attempt to modify the market situation with externalities will give solutions with different, but analogous, externalities.

Campbell, E. Wilson. *An Evaluation of Alternative Land Use and Transportation Systems in the Chicago Area*. Chicago Area Transportation Study. No. 322,021,-I, (October 1968).

Campbell identifies the one land use *alternative* judged to provide the most effective transportation facilities at least cost, for the Chicago area.

——. "Social and Economic Factors in Highway Location." *Journal of the Highway Division American Society of Civil Engineers* 92 (October 1966): 35-48.

Need to develop an evaluative system which considers intangibles. He outlines the Chicago Area Transportation System method of route evaluation which takes a step in this direction.

Carlin, Alan and Wohl, Martin. "An Economic Reevaluation of the Proposed Los Angeles Rapid Transit System." RAND Corp. No. P-3918 (September, 1968).

Criticizing estimates given by Southern California Rapid Transit District in their *Final Report*, they discuss redistribution of real income. They feel that most citizens will find that real costs are greater than the real benefits. Specifically, (1) traffic estimates are highly overoptimistic, and (2) poor cost-benefit procedures were used, so that benefits were overstated relative to costs.

Cervantes, Alfonso J. "To Prevent a Chain of Super-Watts." *Harvard Business Review* 45 (September-October 1967): 55-65.

With respect to unemployment, he feels that industry must use creative leadership and resource potential since he says most of the unemployed are presently unemployable. Therefore, jobs need to be broken down into simpler sequences, and firms need to give training, provide transportation, etc.

Chapin, F. Stuart, Jr. "Activity Systems as a Source of Inputs for Land Use Models." Highway Research Board, National Academy of Sciences, Washington, D.C., Special Report 97, 1967, 77-101.

Location behavior must be integrated with the behavioral system. Land use depends upon activity patterns.

Chenery, Hollis B. and Kretschemer, Kenneth S. "Resource Allocation for Economic Development." *Review of Economics and Statistics* 38 (October 1956): 365-99.

They develop a model based on linear programming and input-output techniques to aid in forming development programs for underdeveloped areas.

Chicago Area Transportation Study. *The Stokie Swift: A Study in Rapid Transit*. No. 317, 031-VI, (July 1968).

Concerned with transit ridership and diversion from other mode. A description of methodology with tables, maps, charts, etc.

Cline, Marvin. "Urban Freeways and Social Structure: Some Problems and Proposals." Highway Research Board, National Academy of Sciences, Washington, D.C., Highway Research Record No. 2, 12-20.

Describes the need for an evaluation method that considers the effect that a change in the physical environment has on the social aspects of an area. He outlines the sources of negative effects and lists a set of corrective proposals.

Coburn. T.M., Beesley, M.E. and Reynolds, D.J. *The London-Birmingham Motorway: Traffic and Economics*. Road Research Laboratory Technical Paper. No. 46, D.S.I.R., J.M.S.O., 1960.

Crumlish, Joseph D. *Notes on the State-of-the-Art of Benefit-Cost Analysis as Related to Transportation Systems*. U.S. Department of Commerce, National Bureau of Standards. Technical Note No. 294, November, 1966.

Although this is based on 1965 (and prior) references, this is a good summary of benefit-cost work done to that time, pointing out the major deficiencies in the method and in the use of it, its possibilities and its limitations. He discusses transportation studies in general, benefit-cost analysis as applied to water resource development studies and highways, gaps in knowledge, lack of agreement on a set procedure, problems with identification and measurement of costs and benefits, data limitations, weighting, intangibles, uncertainty and risk involved with alternatives, etc.

Dalkey, Norman C. *The Delphi Method: An Experimental Study of Group Opinion*. Memorandum RM-5888-TR. Santa Monica, California: The Rand Corporation, 1969.

Dansereau, H.K. "Highway Development: Attitudes and Economic Climate." Highway Research Board, National Academy of Sciences, Washington, D.C., Highway Research Record No. 187, 21-32.

He analyzes attitudes studies, concerning attitudes toward local highway development, attitudes toward planning and zoning practices, and attitude changes toward both development and the planning and zoning practices.

De Alessi, Louis. "Implications of Property Rights For Government Investment Choices." *American Economic Review* 59 (March 1969): 13-24.

Decisions based upon benefit-cost analysis in the absence of property rights are biased since they will depend upon the decision maker's own preference function. Alternatives which affect his own utility maximizing function will be given added weight. This helps to explain why heavy capital investment projects are provided for the future at a rate of discount lower than that of the free market.

Diesing, Paul. "Non-Economic Decision Making." *Ethics* 56 (1955-6): 18-35.

He characterizes noneconomic decision making as a process of integration and adjustment which reduces tension and strain. He analyzes three principles which should be followed in this method.

____. "Socioeconomic Decisions." *Ethics* 69 (1958): 1-18.

Economic decisions allocate resources which sociological decisions reduce tension and strain. He discusses each type of decision making separately and tries to show how they can be applied jointly.

Dodge, William H. *Influence of a Major Highway Improvement upon the Economy of Dunn and St. Croix Counties, Part I, The Local Economy Before Major Highway Improvement*. Wisconsin Commerce Papers. University of Wisconsin, 1962.

Downs, Anthony. "The Law of Peak-Hour Expressway Congestion." *Traffic Quarterly* 16 (July 1962): 393-409.

The cause of peak-hour congestion is not poor planning but the operation of traffic equilibrium. "On urban commuter expressways, peak-hour traffic congestion rises to meet maximum capacity." He sets up a model and discusses several cases, such as the city with car and bus commuters, etc.

He concludes that we need a balanced urban transportation system, and that there is probably not much that can be done about present congestion.

____. *Uncompensated Non-Construction Costs which Urban Highways and Urban Renewal Impose Upon Residential Households*. Real Estate Research Corporation, N.Y. 1970.

This report is a well-written attack on the injustices imposed, particularly on low income, minority-group households, through uncompensated public costs. Downs discussed the types of losses which accrue to households, the compensation principle and desirable modifications of it, and tests which losses must pass to be directly compensable.

He gives rough estimates of the magnitude of some of these uncompensated losses which could be compensated, and concludes that the present system may lead to a misallocation of resources toward highways and urban renewal since displaced households and others nearby are forced to bear 14-21 percent of the real acquisition costs.

Eckenrode, Robert T. "Weighting Multiple Criteria." *Management Science* 22 (November 1965): 180-92.

Tries to develop methods of collecting and weighting data on the relative value of a group of factors relevant to a selected system. Six ranking schemes were used to arrange the data by a team of experts qualified in the field. The results showed general agreement on the selection of the most important criteria and the relative positions of the others.

Edin, Nancy J. *Residential Location and Mode of Transportation to Work: A Model of Choice*. Chicago Area Transportation Study No. 311, 012-VI, October 1966.

She hypothesizes that the journey to work is important in determining residential location since work-trip expenditures form a large part of the budget and since time is important to the commuter. Also, the individual considers location rents and the mode of transport in the trade-off between housing expenditures and transportation expenditures. After reviewing the literature, she sets up a model using a modification of Kain's consumer choice model.

A sequence of behavior choice is assumed: (1) residential density desired by household, (2) whether to purchase a car or not, (3) whether to drive to work or take transit, (4) choice of location of residence (length of journey to work). Skokie is used for empirical part.

Ellis, Raymond. "Review of the Literature on Social Impacts of Transportation Policies." *Strategies For the Evaluation of Alternative Transportation Plans–Part III*, vol. 2, Appendix C. National Cooperative Highway Research Program. Highway Research Board. Project 8-4, Washington, D.C., July 1967.

An outline of the noise, aesthetic, and relocation aspects of highway construction. He criticizes two types of indexes used to measure these factors. Both the economic impact studies and the neighborhood effects studies fail to show these impacts completely.

_____. "Toward Measurement of the Community Consequences of Urban Freeways." Highway Research Board, National Academy of Sciences, Washington, D.C., Highway Research Record No. 229, 1968, 38-52.

He feels that a measure of community impact should consider both the social interaction and spatial aspects of an area. An index of community linkages showing the lines of communication in a region fulfills this dual criteria. By viewing the extent to which highways disrupt these linkages, we can indicate their noneconomic impact.

_____, and Worrall, Richard D. "Toward Measurement of Community Impact: The Utilization of Longitudinal Travel Data to Define Residential Linkages."

Highway Research Board, National Academy of Sciences, Washington, D.C., Highway Research Record No. 277, 1969, 25-39.

This article describes an empirical analysis aimed at defining residential linkages. Determining actual linkages depends upon the evaluation of both household activity patterns and the set of destination points. Using data on residents in Skokie, Illinois a discriminant iterative analysis is undertaken to uniformly apply the linkage criteria to all activity patterns. The results help to indicate the potential community effect of transportation projects.

English, J. Morley ed. *Cost-Effectiveness: Economic Evaluation of Engineered Systems*. New York: John Wiley and Sons, Inc., 1968.

Cost effectiveness is essentially engineering economics (systems engineering) concerned with estimation of costs and evaluation of the worth or effectiveness of systems.

It considers the interrelationships which tie a system together. The emphasis is on comparison of alternatives. Probability theory and decision theory are important, and value theory helps establish worth. He gives a standardized approach to cost effectiveness evaluation (ten steps). He also warns against C-E fallacies and misconceptions (SST is used as a specific example).

Euston, Andrew F. "Design Concepts for the Future." Highway Research Board, National Academy of Sciences, Washington, D.C., Highway Research Record No. 220, 1968, 5-10.

In order to consider all relevant social and economic factors, it is necessary to have a decision-making team representing political, community and professional interests. He further discusses urban design methodology and lists six attempts at building economic and social criteria into the evaluation process.

Evans, Henry K. "Economic Studies for Highways." *Traffic Quarterly* 22 (October 1968): 479-95.

He advocates that more emphasis be given to industry and resource allocation effects in calculating the cost-benefit ratio. At present, such ratios are too concerned with user effects.

Fagin, Henry. "The Penn-Jersey Transportation Study: The Launching of a Permanent Regional Planning Process." *Journal of the American Institute of Planners* 29 (February 1963): 9-18.

Discusses the difficulties encountered in establishing a study of this type. He integrates the Penn-Jersey study with other attemps of this type and describes the problems of alternative evaluation.

Feldstein, Martin S. "Opportunity Cost Calculations in Cost-Benefit Analysis." *Public Finances* 19, no. 2 (1964): 117-39.

Attempts at evaluating the social opportunity cost of funds used in the public sector by discounting at the social time preference rate have been done incorrectly. Most have confused the issue or used incorrect surrogate values.

Fellman, Gordon. "Neighborhood Protest of an Urban Highway." *Journal of the American Institute of Planners* 35 (March 1969): 118-22.

Describes the protest in Cambridge, Mass. over the destruction of low and middle income houses for the "Inner Belt" expressway. Lack of distribution of accurate information led to community fears. Both sides appeared at fault since planners forgot community feelings while citizens forgot highway needs.

——, and Rosenblatt, Roger. "The Social Costs of an Urban Highway: Cambridge and the Inner Belt Road." Transportation and Poverty Conference, June 7, 1968, American Academy of Arts and Sciences, Brookline, Massachusetts.

This is a summary of a study of peoples' attitudes toward the road, their plans, hopes, and fears in connection with the impending highway, their participation or lack of it in a protest movement. It deals only with households actually to be moved for construction of the highway.

Ferguson, C.E. *Macroeconomic Theory of Workable Competition*. Durham, North Carolina: Duke University Press, 1964.

Ferguson, George A. "Development of Transportation System Alternatives." Highway Research Board, National Academy of Sciences, Washington, D.C., Highway Research Record No. 148, 1966, 1-8.

Describes the development of an ideal transportation system based upon minimizing all transportation costs. The existing system is then adjusted to fit the ideal.

Fishburn, Peter C. "Methods of Estimating Additive Utilities." *Management Science* 13 (March 1967): 435-53.

——. "A Note on Recent Developments in Additive Utility Theories for Multi-Factor Situations." *Operations Research* 14 (November-December 1966): 1143-1148.

Franklin, William D. "Benefit-Cost Analysis in Transportation: The Economic Rationale of Resource Allocation." *Traffic Quarterly* 22 (January 1968): 69-75.

Outlines the argument for the fact that transportation is a collective good. Decisions concerning construction must go beyond a simple appeal to the market. They must be made on the basis of benefits and costs to society as a whole.

Freeman, A. Myrick. "Income Distribution and Planning for Public Investment." *American Economic Review* 57 (June 1967), 495-508.

Introduces income redistribution as a policy goal into criteria for selecting public investment projects. Benefit-cost usually considers only changes in national income as an indication of a change in social welfare, however the person or persons to whom the changes accrue are also important.

Friedlaender, Ann Fetter. *The Interstate Highway System: A Study in Public Investment*. Amsterdam, Holland: North Holland Publishing Co., 1965.

Gendell, David S. *Evaluation of Alternative Transportation Systems*. U.S. Department of Transportation, Federal Highway Administration, Bureau of Public Roads, Washington, D.C., Lecture 92.

This article contains very careful definitions of terms. Gendell is specifically interested in goals, objectives, and evaluative criteria with respect to alternatives to be evaluated. He discusses three evaluation procedures: (1) value profile method, (2) rank-based expected value method, (3) value matrix. He includes an example of the value matrix. Gendell also discusses four methods of engineering economy analysis (relating to user costs and benefits only).

Golden, Jay S. *Land Values in Chicago: Before and After Expressway Construction*. Chicago Area Transportation Study. No. 316,051-VI, October, 1968.

An empirical study which supports the hypotheses that increased land values are associated with increased accessibility resulting from expressway construction and measured by distance to the roadway. The empirical data is for roadway test and control areas for Chicago expressways.

Grant, E.L. and Oglesby, C.G. "A Critique of Some Recent Economic Studies Comparing Alternative Highway Locations." *Highway Research Board Proceedings* 39 (1960): 1-8.

A general critique of eighty-five economic impact studies centering on faults that characterize many of them.

____. "Economic Analysis–The Fundamental Approach To Decisions in Highway Planning and Design." *Highway Research Board Proceedings* 39 (1958): 45-57.

An outline of market tests which can be used to justify a particular project. In discounting they advocate the use of the internal rate of return.

____. *Economic Studies for Highways*. Highway Research Board, National Academy of Sciences, Washington, D.C., Bulletin 306, 1961, 23-38.

A description and comparison of five ways of evaluating for dollar costs and benefits. They point out the need for sensitivity analysis, comparison based on marginal benefit-cost ratios and the superiority of the internal rate of return.

Greenhut, Melvin L. "When is the Demand Factor of Location Important?" *Land Economics* 40 (May 1964): 175-84.

He argues that minimization of the cost of production and delivery is inadequate; firms must also consider the demand for the product, and where this demand is located.

Gronau, Reuben and Alcaly, R.E. "The Demand for Abstract Transport Modes: Some Misgivings." *Journal of Regional Science* 9 (April 1969): 153-57.

This is a critique of the Quandt and Baumal article (*Regional Science* 6 (1966): 13-26) which discusses an abstract transportation mode with respect to individual decisions.

Haig, Robert M. "Toward an Understanding of the Metropolis." *Quarterly Journal of Economics* 80 (February 1966).

Hansen, W. Lee and Tiebout, C.M. "An Inter-sectoral Flows Analysis of the California Economy." *Review of Economics and Statistics* 45 (November 1963): 409-18.

In this article, they have reported on their input-output model designed to determine the impact of final demand sectors on employment (direct employment effect and linked employment effect).

Harris, Britton. "Quantitative Models of Urban Development: Their Role in Metropolitan Policy-Making." In *Issues in Urban Economics*, edited by H.S. Perloff and L. Wingo, Jr. Baltimore: Johns Hopkins Press, 1968.

Harris discusses types of models according to classifications by (1) descriptive vs. analytic, (2) holistic vs. partial, (3) macro vs. micro, (4) static vs. dynamic, (5) deterministic vs. probabilistic, and (6) simultaneous vs. sequential.

Haworth, Lawrence. "Deprivation and the Good City." In *Power, Poverty, and Urban Policy*, edited by W. Bloomberg and H. Schmandt. Beverly Hills: Sage Publications, 1968.

He discusses poverty definitions, the good city, and their relationship. Also, he is concerned with the just city, and the response to rioting.

Hegeland, Hugo. *Multiplier Theory*. Lund: C.W.K. Gleerup, 1954.

This work includes an extensive review of the literature and the development of multiplier theory, especially with respect to Keynes and Kahn. Problems occur with regard to the marginal propensity to consume, lags, and leakages. Hegeland concludes that multiplier theory loses relevance when applied to social aggregates; but that it can be fruitfully applied to micro problems (i.e., when applied to the effects of changes in a single stream of expenditure within a limited sector of the economy.

Heenan, G. Warren. "The Economic Effect of Rapid Transit on Real Estate Development." *The Appraisal Journal* 36 (April 1968): 213-24.

Sees rapid transit as the base of a balanced transportation system in an urban area, as creating and maintaining urban land values. Toronto experience.

———. "The Impact of Rapid Transit on Business and Real Estate in the Central City." Speech to Market Street Development Project. San Francisco, March 30, 1967.

Basis for later paper. Author believes in federal support of a balanced metro. transportation system, feels public transportation should be free, like education.

Henderson, P.D. "Notes on Public Investment Criteria in the United Kingdom." *Bulletin of the Oxford University Institute of Economics and Statistics* 27 (February 1965).

A survey of the literature concerning the application of principles of benefit cost analysis to the area of public enterprise investment decisions.

Hennes, Robert G. "Highways As an Instrument of Economic and Social Change." Highway Research Board, National Academy of Sciences, Washington, D.C., Special Report 56, 1960, 131-35.

An analysis of the nonuser importance of major highway construction. In the earlier periods of road paving the social impact of highways was recognized. However the introduction of economic analysis based on user benefits and costs combined with planning geared to user origin and destination surveys caused the social impact to be neglected. Today, any major highway that is contemplated must include this social impact in the decision making process.

Herfindahl, Orris C. and Kneese, Allen V. *Quality of the Environment.* Baltimore: Johns Hopkins Press, 1965.

They deal with problems connected with water pollution, air pollution, chemicals as pesticides, the physical environment of urban places, and use of rural areas.

Hicks, John R. "The Foundations of Welfare Economics." *Economic Journal* 49 (September 1939): 696-712.

Highway Research Board. Division of Engineering. "Concepts, Procedure, and Evaluation Pertaining to Economic Analysis of the Economic and Social Consequences of Highway Improvements." Project 2-11, Draft Report. National Cooperative Highway Research Program, Highway Research Board. Washington, D.C., (January 1969).

A complete study of all current information on the existence and measurement of social and economic effects of highways. It covers both the evaluation process and the actual method of decision making.

_____. *Highway Capacity Manual, 1965.* National Academy of Sciences, Washington, D.C., Special Report 87, 1965.

This is a technical report, concerning traffic characteristics, capacity and level of service, intersections, weaving, ramps, freeways, streets, bus transit.

_____. Special Report No. 69. *A Key to Change, Urban Transportation Research.* National Research Council, Washington, D.C., 1962.

This report shows concern with the total impact of roads on changing land values, reorganization of land use, and changes in employment and the labor supply. It stresses the interrelationship of transportation and economic development in a region, and the multi-dimensionality of public decision making. Research projects are proposed.

Hill, Morris. "A Goals-Achievement Matrix for Evaluation of Alternative Plans." *Journal of the American Institute of Planners* 34 (January 1968): 19-29.

Hill defines rational planning as a process for determining future action by utilizing scarce resources in such a way as to maximize the attainment of a set of given ends. Again he criticizes cost-benefit analysis, and discusses his goals-achievement matrix, with respect to quantitative and qualitative objectives. The major disadvantage of the analysis is that interaction and interdependence between objectives is not considered; therefore, the analysis is limited to use for plans in a single sector rather than for multi-sector planning.

_____. "A Method for the Evaluation of Transportation Plans." Highway Research Board, National Academy of Sciences, Washington, D.C., Highway Research Record No. 180, 1967, 21-34.

Hill criticizes cost-benefit analysis as being suitable only for the evaluation of the economic efficiency of alternative transportation projects. He proposes goal achievement analysis as an alternative method, where the planner is interested in the entire set of objectives in a single system. He suggests a weighting system to be applied to the objective, subgroups, sectors, locations, and activities affected, and set up a hierarchy of goals in an attempt to deal in operational forms.

Hill, Stuart L. "The Effects of Freeways on Neighborhood." Right of Way Research and Development Division of Highways, Dept. of Public Works Highway Transportation Agency, State of California. In cooperation with the U.S. Department of Transportation Federal Highway Administration Bureau of Public Roads, June, 1967.

He develops a "mobility index" to establish the location and boundary of neighborhood. The index reflects the cultural and behavioral aspects of the area by indicating the stability of residences and the propensity to cultural change.

____. "Watts-Century Freeway." Mimeographed paper presented at the Transportation and Community Values Conference. Highway Research Board, 1969.

A description of how the California Highway Department has tried to use the construction of a freeway through Watts as a means of bettering the living standards of the entire area. Through the action of the legislature and the department, replacement housing was made available and it was purchased by those displaced by the freeway, using the right of way acquisition money.

____, and Frankland, Bamford. "Mobility as a Measure of Neighborhood." Highway Research Board, National Academy of Sciences, Washington, D.C., Highway Research Record No. 187, 1967, 33-42.

A mobility index is developed to assist in defining a neighborhood.

Hines, Lawrence G. "The Hazards of Benefit-Cost Analysis as a Guide to Public Investment Policy." *Public Finance* 17, no. 2 (1962): 101-117.

A complete outline of many of the problems which can arise when benefit-cost analysis is applied in analyzing public investments.

Hinkle, Jere J. and Frye, Frederick F. *The Influence of an Expressway on Travel Parameters–The Dan Ryan Study*. Chicago Area Transportation Study. No. 10, 320-VI, June 1965.

This is a before and after study using roadside interviewing and handout postcard techniques. They studied traffic redistribution by volumes, changes in accident rates, diversion by facility, change in trip length. They conclude that additional traffic capacity leads to traffic redistributing itself by upgrading, which implies a lower accident rate, lower vehicle operating costs, etc. An interesting term is defined: "adverse travel" is the additional distance that a person will travel in return for a benefit.

Hinrichs, Harley H. and Taylor, Graeme M. (eds.). *Program Budgeting and Benefit-Cost Analysis*. Pacific Palisades, California: Goodyear Publishing Co., 1969.

Hirsch, Werner Z. "Design and Use of Regional Accounts." *American Economic Association* (May, 1962).

Hirsch explains the use of accounts for regional and intraregional decisions, and discusses the types of data needed.

———, ed. *Elements of Regional Accounts*. Baltimore: Johns Hopkins Press, 1964.

Papers by Henderson, Burkhead, Niskanen, Harris, Schnore, Stolnitz, Perloff, and Levin make up a good background reader in regional accounts, presenting varying ideas.

———. "Expenditure Implications of Metropolitan Growth and Consolidation." *Review of Economics and Statistics* 41 (August 1959): 232-41.

What are the likely expenditure effects of metropolitan growth and consolidation. He discusses horizontally and vertically integrated services, concluding that public education, fire and police protection, refuse collection, etc., are horizontally integrated, so that growth and consolidation will have little effect on per capita expenditure levels. Central administration of municipal or district governments tends to be circularly integrated, so that per capita expenditures will fall in early growth, then rise as the community passes a medium size. Water and sewage services are vertically integrated, so that growth and consolidation imply falling per capita expenditures until a very large scale is reached. Thus, efficiency considerations do not imply across-the-board consolidation of metropolitan areas governments.

———. "Input-output Techniques for Urban Government Decisions." *Papers and Proceedings of the American Economic Association*. (May 1968), pp. 162-70.

He presents an urban public services expenditure projection model and a regional planning model to evaluate plans in terms of their net fiscal health results. He treats urban government as a closed regional input-output model, and disaggregates urban government services into five sectors. In sum, the model is I-O with subsidiary calculations. He does some empirical work with St. Louis data, but feels that the results are not very satisfactory.

———. "Interindustry Relations of a Metropolitan Area." *Review of Economics and Statistics* 41 (August 1959): 360-69.

Hirsch estimates local technical coefficients for St. Louis and then sets up an interindustry flow table in order to predict the local income multiplier effect from new household income generation. This input-output report includes results for income and employment multipliers, exports, and area stability. The model uses gross, rather than net, flows.

Hirshleifer, J., De Haven, J.C. and Milliman, J.W. *Water Supply Economics, Technology and Policy*. Chicago: University of Chicago Press, 1960.

Hitch, C.J. and McKean, R.H. *The Economics of Defense in the Nuclear Age*. Cambridge: Harvard University Press, 1961.

Hochwald, Werner, ed. *Design of Regional Accounts*. Baltimore: Johns Hopkins Press, 1961.

This book includes papers by Hirsch, Ruggles and Ruggles, Levin, Delwart and Sonenblum, Berman, Borts and Stein, and Perloff. It is from one of the earlier conference on regional accounts.

Hoover, Edgar M. "The Evolving Form and Organization of the Metropolis." In *Issues in Urban Economics*, edited by H.S. Perloff and L. Wingo, Jr. Baltimore: Johns Hopkins Press, 1968.

This paper is a review of the state-of-the-arts for understanding the economic forces affecting the spatial pattern of activities in urban areas. He identifies basic location factors, considers empirical studies of spatial patterns, and discusses changes and adjustment problems (with implications for urban planning).

——, and Chinitz, Benjamin. "The Role of Accounts in the Economic Study of Regions." In *Design of Regional Accounts*, edited by Werner Hochwald. Baltimore: Johns Hopkins Press, 1961.

They define accounts as systematic and quantitative cross-tabulations of economic transactions and claims, primarily in money units and primarily cross-sectional. (They do not consider balance-sheet accounting.) They discuss exogenous vs. endogenous emphasis in regional economic analysis, and feel that the horizontal (endogenous) approach may be the most feasible due to fund problems.

Irwin, Neal A. "Criteria for Evaluating Alternative Transportation Systems." Highway Research Board, National Academy of Sciences, Washington, D.C., Highway Research Record No. 148, 1966, 9-12.

Describes four classifications of criteria used to evaluate alternative plans. He favors the use of minimum standards of acceptance where intangibles are involved with relative criteria being used to compare those plans which pass the minimum test.

Jessiman, William; Brand, David; Tumminia, Alfred; and Brusse, C. Roger. "A Rational Decision Making Technique for Transportation Planning." Highway Research Board, National Academy of Sciences, Washington, D.C., Highway Research Record No. 180, 1967, 71-80.

Employs a form of utility analysis to evaluate alternatives. Weighted objectives are used to estimate the total utility of each plan, by multiplying the weight times the value each plan has in achieving the objective.

Kain, John F. "The Journey to Work as a Determinant of Residential Location." *Papers and Proceedings of the Regional Science Association*. 1962, pp. 137-60.

Kain hypothesizes that households substitute journey-to-work expenditures for site expenditures, with the substitution dependent primarily on household preferences for low-density as opposed to high-density residential services. He set up a model for the locational choice of a single household, where transportation costs, in terms of money and time, rise with the distance of the residence from the workplace. In the market for residential space, price is assumed to vary with location, so that rents fall with increasing distance from the job.

——. *A Report on an Urban Transportation Model, Some Progress and Some Problems*. RAND, No. P-2549, Santa Monica, Calif., June, 1962.

Kain sets up a model of a community to help specify the relevant demand parameters for urban transportation. The model is a series of statistically estimated submodels tied together by machine instructions (decision rules).

He feels importance of secondary effects has been badly overestimated, but any partial analysis should indicate the nature and extent of these secondary effects. The RAND model is designed to study the feedback between the provision of transportation facilities and changes in land use. It concentrates on explaining the locational behavior of firms and households, using journey-to-work travel as the major variable for urban transportation planning.

It is a useful summary of the two earlier RAND works.

——, and Meyer, John R. "Computer Simulations, Physio-economic Systems, and Intraregional Models." *American Economic Review* 57 (May 1968): 171-81.

They discuss computer simulations and their role in the evaluation of large-scale public investment programs, their application to urban and regional problems, problems and deficiencies of metropolitan growth models, and data requirements.

——. *A First Approximation to a RAND Model for Study of Urban Transportation*. RM-2878-FF, RAND Corp., Santa Monica, Calif., November, 1961.

This is a *preliminary* work on studying the intricate relationships between transportation and the spatial organization of economic activities (similar to earlier RAND paper of 1961). It considers three sectors: industrial, commercial, and residential. This is an econometric model, based on relationships estimated from cross-sectional data. It is not a forecasting tool, but aims at the general urban area rather than at evaluating the socioeconomic effect of alternative transportation proposals.

——, and Wohl, M. *The Urban Transportation Problem*. Cambridge: Harvard University Press, 1965.

Kaldor, Nicholas. "Welfare Propositions of Economics and Interpersonal Comparisons of Utility." *Economic Journal* 49 (September 1939): 549-52.

Initial formulation of the compensation principle of welfare theory.

Kanwit, Edward L. and Eckartt, Alma F. "Transportation Implications of Employment Trends in Central Cities and Suburbs." Highway Research Board, National Academy of Sciences, Washington, D.C., Highway Research Record No. 187, 1967, 1-14.

Lambrakis, Helen C. *A Report on Community Socio-Economic Analysis for the Crosstown Expressway Corridor*. November 30, 1964.

She feels that the process of social selection and segregation which tend to create natural social groups also tend to determine the "natural areas" of the city. She set out to define the natural areas (clustering of similar activities and population) and the predominant social economic characteristics in each of these areas for parts of Chicago. The report shows the variables and research techniques used. This type of work may be useful for helping to define communities.

Lampard, Eric E. "The Evolving System of Cities in the United States: Urbanization and Economic Development." In *Issues in Urban Economics*, edited by H.S. Perloff and L. Wingo, Jr. Baltimore: Johns Hopkins Press, 1968.

Lampard reviews the literature, then analyzes data to explain the directions of urban change.

____. "The History of Cities in Economically Advanced Areas." *Economic Development and Cultural Change* 3 (January 1955): 81-136.

This is a general article about the evolution of cities, with respect to urban growth. It presents the prevailing theories about the growth of cities.

Lange, A.S. and Wohl, Martin. "Evaluation of Highway Impact." Highway Research Board, National Academy of Sciences, Washington, D.C., Bulletin 268, 1960.

They feel that the economic aspects of highway impact studies are poorly done since nonuser benefits are only transferred user benefits. Studies centering only on nonuser benefits do not provide the information needed to make location and design decisions.

Leontief, Wassily W. "Input-Output Economics." *Scientific American* 185 (October 1951): 15-21.

This article explains input-output analysis and shows a portion of the 1947 table for the exchange of goods and services in the United States.

____. "The Structure of the U.S. Economy." *Scientific American* 252 (April 1965): 25-35.

Leontief refined his earlier work on input-output tables; this article explains some of the changes that have been made and some of the uses to which I-O tables have been put.

Levin, Charles L. "Establishing Goals for Regional Economic Development." *Journal of the American Institute of Planners* 30 (May 1964): 100-109.

Levin relates the usual goal of maximization of regional output to models containing multiple development goals, and also to regional and national goal conflicts which may occur. However, unless the development goals are specified beforehand, his models do not indicate an optimum policy.

____. "Regional and Interregional Accounts in Perspective." *Papers and Proceedings of the Regional Science Association*. 1964, 127-44.

Levin reviews the development of regional accounts, discusses measurement problems, conceptual and methodological problems. He presents his foreign-trade employment and value added multipliers. In conclusion, he points out that there is no correct way of setting up regional accounts independently of the preconceived analytical purpose to which they are to be put. Thus, definitions and measures will depend on the purpose of the accounts.

____. *Theory and Method of Income and Products Accounts for Metropolitan Areas*. Pittsburgh: University of Pittsburgh, 1963.

Case study of the Elgin-Dundee area, he analyzes methods used in most regional economic base studies. He includes the methodology used in this particular study which was designed to measure the relative contribution of the several industries in a community to the area's total economic activity. He discusses defining the target area, the survey, and research questionnaires, the actual estimated magnitudes for product accounts of the area, and he computes the foreign trade value-added multiplier.

Lichfield, Nathaniel. "Cost-Benefit Analysis in City Planning." *Journal of the American Institute of Planners* 26 (November 1960): 273-79.

Views the impact of alternative plans in dollar terms through the use of social accounting. The resulting "Planning Balance Sheet" divides the community by the groups affected and measures dollar impact through dollar costs and benefits.

____. "Economics in Town Planning." *Town Planning Review* 38 (April 1968): 5-20.

Review of his "Planning Balance Sheet" system comparing it to other plans and adding that the results of the "Sheet" must be compared to the communities goals to properly view potential impact.

Losch, August. "The Nature of Economic Regions." *Southern Economic Journal* (August 1963).

Beginning with an undifferentiated plan, Losch hypothesizes that the demand cone due to shipping costs from the center of production will lead to hexagonal market areas as the most economical shape for trading areas. Then nets of these hexagonal areas will develop for different products, etc. This is one of the first articles in location theory.

Lowry, Ira S. *A Model of Metropolis*. RAND Corp. RM-4035-RC, Santa Monica, Calif., August, 1964.

His computer model of the spatial organization of human activities within a metropolitan area may be used to (1) evaluate the impact of public decisions on the metropolitan form, and (2) to predict changes in the metropolitan form over time as a consequence of changes in key variables (such as the pattern of basic employment, the efficiency of the transportation system, growth of population). This location model is designed to assign urban activities to subareas of a bounded region. The model considers three kinds of activities: the basic or export sector (exogenous); the retail, or residual or local-oriented sector (endogenous); and the household sector (endogenous).

____. "Seven Models of Urban Development: A Structural Comparison." Highway Research Board, National Academy of Sciences, Washington, D.C., Special Report 97, 1968.

Lowry sets up a model of the market for urban land, then compares seven models to this.

1. Land use–Chicago Area Transportation Study Model.
2. Land use succession–University of North Carolina Model.

3. Location–Empiric Model.
4. Migration–Polimetric Model.
5. Hybrid–Pittsburgh Model.
6. Market demand–Penn-Jersey Model.
7. Market supply–San Francisco Model.

He gives a good summary and classification of these models.

———. "A Short Course in Model Design." *Journal of the American Institute of Planners* 31 (May 1965): 158-66.

In a good review article, he discusses descriptive, planning and predictive models, and the uses of, requirements for, and limitations of each. He talks about aggregation and the time element, and the concept of change. Four solution methods are given with respect to fitting, testing, and evaluation of a model.

Maass, Arthur. "Benefit-Cost Analysis: Its Relevance to Public Investment Decisions." *Quarterly Journal of Economics* 80 (May 1966): 208-26.

Description of current practices and areas of future need in the application of benefit-cost analysis to analyze public investments. In particular, a set of trade-offs between economic efficiency and other criteria is needed.

Manheim, Marvin L. "Principles of Transport Systems Analysis." Highway Research Board, National Academy of Sciences, Washington, D.C., Highway Research Record No. 180, 1967, 11-20.

He discusses nine principles which can be used as guides both in viewing problems and in their analysis. Five of the principles consider what must be incorporated into a transport system and what interactions will result while the other four show and outline the available alternatives and factors to be considered in decision making.

MacRae, Duncan. "Economic Evaluation of Urban Development." MIT, preliminary draft of a research paper, Spring, 1969.

Assuming that social goals are given, MacRae proposes an alternative to benefit-cost analysis in the form of his theory of social policy. With a given quadratic social preference function on the deviation between goals and actual levels of variables, the object is to choose policy variables to minimize the value of the social preference function. He applies the model using the EMPIRIC location model.

Mann, Patrick. "The Application of Users Charges for Urban Public Services." *Reviews in Urban Economics* (Winter 1968): 25-46.

In this review article about pricing public services, Mann concludes that pricing of these services should be considered with respect to the large urban picture. He advocates experimentation with approximations to marginal cost pricing to try to find more rational pricing schemes for urban public services.

Marglin, Stephan A. "The Opportunity Costs of Public Investment." *Quarterly Journal of Economics* 77 (May 1963): 274-89.

———. "The Social Rate of Discount and the Optimal Rate of Investment." *Quarterly Journal of Economics* 77 (February 1963): 95-111.

Meier, Richard L. "Human Time Allocation: A Basis for Social Accounts." *Journal of the American Institute of Planners* 25 (February 1959): 27-33.

Alterations in the existing activity pattern of time allocation can indicate the direction of community welfare change brought about by construction of a highway. However, such a measure may not be as sensitive to cultural change as planners would like.

———. "Measuring Social and Cultural Change in Urban Regions." *Journal of the American Institute of Planners* 26 (November 1959): 180-90.

The impact of change is best analyzed by viewing alterations in communications. Alterations in social transactions and the institutions through which they take place can give an indication of cultural change. Impact is measured by changes in the direction and rates of flow of social communications.

Meyer, John R. "Regional Economics: A Survey." *American Economic Review* 53 (March 1963): 9-54.

Meyer understands the definitional problems with any type of regional analysis, and gives a definition of an economic development region. He discusses the theoretical foundations of regional economics, and types of approach to regional studies. His main criticism is that most of the behavioral hypotheses have not been adequately formulated and tested. The bibliography given is extensive.

Myerson, Margy E. "Implications of Sociological Research for Urban Passenger Transportation." *Highway Research Board Proceedings*. Vol. 34, 1955, pp. 1-7.

Miernyk, William H. *The Elements of Input-Output Analysis*. New York: Random House, 1967.

In this introductory treatment, Miernyk discusses application of input-output analysis to regional and interregional studies, and to international developments. He also explains the mathematics necessary to work with this type of analysis.

Mohring, Herbert D. "Urban Highway Investments." In *Measuring Benefits of Government Investments*, edited by Robert Dorfman. The Brookings Institution, 1965.

Use of consumer surplus and pricing techniques to measure the benefits of highways. He relates the theoretical analysis to actual data for the twin-cities area of Minnesota.

———, and Harwitz, Mitchell. *Highway Benefits: An Analytical Framework*. Evanston, Ill.: Northwestern University Press, 1962.

They estimate the benefits on existing use and the substitution benefits of a highway improvement by viewing the area under the demand curve for the transportation facility.

Moore, Frederick T. "Operations Research on Urban Problems." RAND Corporation, No. P-3414, Santa Monica, Calif., June, 1966.

Moore discusses the use of O-R in making the trade-off between spending for public goods (such as between health and transportation expenditures). The main problems are (1) what are appropriate criteria for choosing among alternative programs (evaluating nonmarket goods and services) and (2) what incentives can be designed to make a synthetic market work reasonably well as an alternative to direct government intervention. He stresses the importance of the total picture in decision making.

____, and Petersen, James W. "Regional Analysis: An Interindustry Model of Utah." *Review of Economics and Statistics* 37 (November 1955): 368-83.

The interindustry model is a complete structural description of the transactions occurring in a region (arbitrarily defined) during a base period. The model is similar to a Leontief model, but takes account of differences between demand for goods satisfied for local production and demand satisfied by imports. Although the assumptions are somewhat restrictive, the model is useful for Utah since Utah has a space buffer to isolate it.

Morehouse, Thomas A. "The 1962 Highway Act: A Study in Artful Interpretation." *Journal of the American Institute of Planners* 35 (May 1969): 160-68.

Moses, Leon N. "Towards a Theory of Intra-Urban Wage Differentials and their Influence on Travel Patterns." *Papers and Proceedings of the Regional Science Association* 9 (1962): 53-64.

Under his assumptions, net income depends on the structure of intraurban transport costs and the distance of the residence from the city core. For auto transport, the cost of overcoming distance is assumed to rise as the core is neared since congestion rises. Thus, those living nearer the job enjoy location rents since he assumes a uniform wage sufficient to draw workers from farther points. However, he does not consider population distribution, and he assumes the wage is given rather than determined.

____, and Williamson, Harold F. "Value in Time, Choice of Mode, and the Subsidy Issue in Urban Transportation." *Journal of Political Economy* 71 (June 1963): 247-64.

McKain, Walter C. "Community Response to Highway Improvement." Highway Research Board, National Academy of Sciences, Washington, D.C., Highway Research Record no. 96, 1965, 19-23.

A report based on a study of the Connecticut Turnpike which indicated that the construction of a highway can have significant area impacts if the capacity for accepting regional change exists.

McKean, R.N. *Efficiency in Government through Systems Analysis*. New York: John Wiley and Sons, 1958.

McMillan, T.E., Jr. "Why Manufacturers Choose Plant Locations vs. Determinants of Plant Locations." *Land Economics* 51 (August 1965): 239-46.

Nash, William W. and Voss, Jerrold R. "Analysing the Socio-Economic Impact of Urban Highways." Highway Research Board, National Academy of Sciences, Washington, D.C., Bulletin No. 268, 1960, 80-94.

An examination of the impact of the Boston "Inner Belt" both in the short-run displacement and in long-run growth effects. A "with and without" study was made showing the peak effects will occur four to eight years after the completion of the system.

National Cooperative Highway Research Program, Report 49. *National Survey of Transportation Attitudes and Behavior, Phase I, Summary Report.* R.K. McMillan and H. Assael. Clinton Research Services, Philadelphia, Highway Research Board, 1968.

Two independent nationwide surveys were conducted to determine whether existing procedures for allocating money for highways are really responsive to public attitudes and behavior relating the transportation of people. Attitudes toward the auto are generally favorable, the value placed on the car is high, and people seem to have more detachment when considering public transportation.

National Cooperative Highway Research Program. *National Survey of Transportation Attitudes and Behavior, Phase II*, Final Report–Project No. HR 20-4. R.K. McMillan and H. Assael. Clinton Research Services, Philadelphia, Highway Research Board, 1968.

Newcomb, Robinson. "A New Approach to Benefit-Cost Analysis." Highway Research Board, National Academy of Sciences, Washington, D.C., Highway Research Record No. 138, 1966, 18-21.

He advocates a with and without form of impact evaluation where the project itself is likely to alter the underlying parameters. The difference between the two estimates serves as a measure of project benefits.

Niedercorn, John H. and Kain, J.F. "Econometric Model of Metropolitan Development." *Papers and Proceedings of the Regional Science Association* 10, 1963, 123-43.

This is a preliminary dynamic cross-sectional model describing population and employment changes in the thirty-nine largest SMSA's from 1954-58. They conclude that secular shifts in manufacturing employment are the most powerful single determinant of variations in the rate of metropolitan growth.

North, Douglas C. "Location Theory and Regional Economic Growth." *Journal of Political Economy* 63 (June 1955): 243-58.

He reviews the literature concerning usual growth sequences, and argues that industrialization is not necessary for the economic growth of a region, and that possiblities for industrialization have been made to seem far more difficult than they really are.

Novick, David. *Program Budgeting: Program Analysis and the Federal Budget.* Cambridge: Harvard University Press, 1965.

A collection of essays outlining the structure, application and usefulness of program budgeting in the federal area.

Oregon State Highway Department. *Technical Bulletin No. 7.* By McCullough, C.B. and Beakey, John. Salem, Oregon, 1938.

Owen, Wilfred. *Strategy for Mobility*. Washington, D.C.: Brookings Institution, 1964.

____. "Transportation and Technology." *American Economic Review* 52 (May 1962): 405-13.

Paine, F.T., Nash, A.N., Hille, S.J. and Brunner, G.A. *Consumer Conceived Attributes of Transportation: An Attitude Study*, prepared for the U.S. Department of Commerce, Bureau of Public Roads, by the University of Maryland Department of Business Administration. June, 1967.

The objective of the study is to develop an evaluation of research methodology which permits the efficient accumulation of consumer transportation attitude information. They attempted to identify and assess the importance of attributes of an ideal transportation system as conceived by the consumer, and to determine the extent to which consumers consider existing systems to satisfy this ideal.

Peacock, A.T. and Doscer, D.G.M. "Regional Input-Output Analysis and Government Spending." *Scottish Journal of Political Economy* 6 (November 1959): 229-36.

Peattie, Lisa R. "Reflections on Advocacy Planning." *Journal of the American Institute of Planners* 34 (March 1968): 80-88.

Planners may help the "people" make their interests felt through overt political planning. The clients of advocacy planners may be difficult to define since the neighborhood may not be a cohesive, agreeing group. Natural clients would be organizations appearing in response to some threat, but they might not be representative of community interests. This new kind of politics may be a way of dealing with issues which are not particularly to any single local community but to a group interest.

Perazich, George and Fischman, Leonard L. "Methodology for Evaluating Costs and Benefits of Alternative Urban Transportation Systems." Highway Research Board, National Academy of Sciences, Washington, D.C., Highway Research Record No. 148, 1966, 59-71.

They divide the community into impact groups and engage in a review of all economic principles use in undertaking a cost-benefit study. They particularly emphasize the need for sensitivity analysis.

Perloff, Harvey S. *How a Region Grows*. Supplementary Paper No. 17. Committee for Economic Development, 1963.

Based on the larger volume, *Regions, Resources, and Economic Growth, 1870-1960*, this work discusses growth factors, long-run changes in the distribution of economic activity, and various approaches to regional development.

____. "A National System of Metropolitan Information and Analysis." *American Economic Association* 52 (May 1962).

Perloff thinks of a metropolitan area primarily as a labor market. He feels that the major needs for accounting systems are (1) metering for ascertaining the state of the community, (2) projections of growth of population and economic activities and of structural changes, (3) impact or interaction analysis, and (4) evaluation of specific programs, projects, and policy alternatives.

____. "Relative Regional Economic Growth: An Approach to Regional Accounts." *Design of Regional Accounts*. Baltimore: Johns Hopkins Press, 1961.

Pfouts, R.W. (ed.). *Techniques of Urban Economic Analysis*. W. Trenton, New Jersey: Chandler Davis Publishing Co., 1960.

Pillsbury, Warren A. "Economics of Highway Location: A Critique of Collateral Effects Analysis." Highway Research Board, National Academy of Sciences, Washington, D.C., Highway Research Record No. 75, 1965.

Three methods of measuring actual highway impact are compared; engineering economy, marginal analysis, and collateral effects analysis. He feels that collateral effects analysis which includes secondary and tertiary benefits and costs is the best means of viewing potential impact.

Prest, A.R. and Turvey, R. "Cost-Benefit Analysis: A Survey." *Economic Journal* 75 (December 1965): 683-735.

A review of the history and major issues of cost-benefit analysis. In outlining the methodology such problems as externalities, choice of discount rate, the prices to be used and possible constraints are discussed. They also survey several studies showing practical applications.

Quandt, Richard E. and Baumol, William J. "The Demand for Abstract Transport Modes." *Journal of Regional Science* 6 (Winter 1966): 13-26.

They define an abstract transport mode by a set of characteristics such as speed, fare, frequency of travel, service, and comfort. Then the decision to travel by a mode is a function of (1) the absolute levels of the characteristics of the "best" modes on each criterion, and (2) the performance level of the mode on each criterion relative to the best mode.

____. "The Demand for Abstract Transport Modes: Some Hopes." *Journal of Regional Science* 9 (April 1969): 159-62.

This is a defense of their 1966 article, including a discussion of later refinements in the theory and in their statistics.

RAND Corporation. "Transportation for Future Urban Communities: A Study Prospectus." RM-28 24-FF, Santa Monica, Calif., 1961.

RAND sees a transportation system as a response to the growth of an area rather than as a causative factor. This work includes a general review of the types of studies done up to 1961, and states that most valuable contribution of previous studies is an estimation of the scale and directional distribution of future trips. It states that more information is needed on the urban structure for jointly determining land use and travel patterns as well as long-run forecasts. The model uses status variables (or community characteristics), input variables (exogenous) and output variables (community characteristics at the end of the period). This is a preliminary statement, defining the directions of study plans in 1961.

Rodwin, Lloyd. "Choosing Regions for Development." In *Public Policy*, edited by Carl J. Friedrich and Seymour E. Harris. Harvard Graduate School of Public Administration. Cambridge: Harvard University Press, 1963.

After reviewing the literature, Rodwin discusses criteria for selecting regions for development efforts in underdeveloped countries, given specified goals (such as a certain minimum rate of increase in per capita income).

Rogers, A.C. "The Urban Freeway: An Experiment in Team Design and Decision-Making." Highway Research Board, National Academy of Sciences, Washington, D.C., Highway Research Record No. 20, 1968.

He describes a new design and decision-making approach involving a team of technical and academic experts who work with all the political, private and community groups affected by the decision. Together they design and locate the highway. Hopefully this will assure consideration of all possible impacts.

Schaller, Howard G. *Public Expenditure Decisions in the Urban Community*. Baltimore: Johns Hopkins Press, 1963.

Schimpeler, Charles C. and Grecco, William L. "The Community Systems Evaluation: An Approach Based on Community Structure and Values." Highway Research Board, National Academy of Sciences, Washington, D.C., Highway Research Record No. 238, 1968, 123-52.

They develop a weighted set of community goals and objectives which are used to evaluate the transportation plan. The weights are expressed in the form of utility estimates while plan effectiveness with respect to each objective is expressed in terms of probability of achievement. The plan can be further disaggregated to consider and weight the individual groups affected.

Sears, Bradford. "Highways as Environmental Elements." Highway Research Board, National Academy of Sciences, Washington, D.C., Highway Research Record No. 93, 1965, 49-53.

He is concerned with the fact that roads were designed for vehicles not people, and environmental or land effects are ignored. The design of the highway must preserve the natural contour of the landscape.

Shaffer, Margaret. "Attitudes, Community Values and Highway Planning." Highway Research Board, National Academy of Sciences, Washington, D.C., Highway Research Record No. 187, 1967, 55-61.

She lists several attempts to identify attitudes and defines them giving useful techniques for measurement. In analysis, attitudes are correlated with socio-economic characteristics to derive the values used to evaluate alternative plans. The weighting and elimination of conflicting values is necessary to ensure the selection of the best highway plan.

Shostal, H. Claude. "Pricing and Planning Urban Transportation." *Review in Urban Economics* (Fall 1968), 5-28.

He is concerned with urban transportation from the point of view of pricing for the present and investing for the future. He contrasts the views presented by Kuhn vs. Meyer, Kain, etc. with regard to planning and marginal cost pricing of existing facilities. Problems of quantification are also considered. In appendices, Kuhn's objectives of public enterprise and model balance sheet for transportation proposals are given, as well as sample cost formula by Meyer, Kain, and Wohl.

Sirkin, Gerald. "The Theory of the Regional Economic Base." *Review of Economics and Statistics* 41 (November 1959): 426-29.

Sonenblum, Sidney. "The Uses and Development of Regional Projections." In *Issues in Urban Economics*, edited by Harvey S. Perloff and Lowden Wingo, Jr. Baltimore: Johns Hopkins Press, 1968.

Sonenblum distinguishes between prediction and planning in discussing why regional information is needed. He also considers the problem of defining a region, and different ways to view regional change. He summarizes techniques of projecting selected variables such as income or population.

Spiegelman, Robert G. "Activity Analysis Models in Regional Development Planning." *Regional Science Association Papers* 17 (1966): 143-59.

He uses activity analysis as an information base for optimization in economic development planning. The planner must specify the objective to be maximized or minimized, alternative plans to be considered, and constraints on the use of resources and specific targets to be met. He discusses dynamic aspects, pointing out that cost may be a major constraint on construction of the model.

_____, Baum, E.L. and Talbert, L.E. *Application of Activity Analysis to Regional Development Planning*. Dept. of Agriculture Technical Bulletin. No. 1339, Washington, D.C., 1965.

This is a case study of the development and utilization of an econometric model for planning economic development in small rural areas. The price solution of the activity analysis planning model directs the type of investment which should be made.

Starr, Richard. *Urban Choices*. Baltimore: Penguin Books, 1967.

Steger, Wilbur A. and Lakshmann, T.R. "Plan Evaluation Methodologies: Some Aspects of Decision Requirements and Analytical Response." Highway Research Board, National Academy of Sciences, Washington, D.C., Special Report 97, 1967, 33-76.

A general review of many of the problems encountered in plan evaluation including issue relevance, process context, plan design, and identification of impacts. The need for trade-offs and the possible inability to aggregate are discussed.

Stoll, Walter D. "Changes in Average Trip Length: A Case Study by Mode and Purpose of Skokie Trips made in 1956 and 1964." Chicago Area Transportation Study, No. 311, 014-VI, December, 1967.

Stolnitz, George J. "The Changing Profile of our Urban Human Resources." In *Issues in Urban Economics*, edited by H.S. Perloff and L. Wingo, Jr. Baltimore: Johns Hopkins Press, 1968.

Steiner, Peter O. "The Role of Alternative Cost in Project Design and Selection." *Quarterly Journal of Economics* 79 (August, 1965): 417-30.

He demonstrates that given a meaningful alternative, all or some of the benefit calculation can be eliminated by viewing the costs of the alternative foregone. This is a reversal of the normal role of opportunity cost.

Thiel, Floyd I. "Social Effects of Modern Highway Transportation." Highway Research Board, National Academy of Sciences, Washington, D.C., Bulletin 327, 1962.

A survey of major highway impact studies centering on their analysis of the sociological consequences. Impacts upon recreation, rural and urban living patterns, public services, and nonwork opportunities are considered.

Thomas, Edwin N. and Schofer, Joseph L. *Strategies for the Evaluation of Alternative Transportation Plans*. Final Report Vols. I, II, III, National Cooperative Highway Research Program Project 20-8. The Transportation Center, Northwestern University, Evanston, Ill.: 1967.

A description of the systems analysis approach in the evaluation and selection of alternative transportation plans. The interaction of the transportation and activity systems is emphasized. The procedural steps used in systems analysis are outlined and it is pointed out that the method is subjective and merely organizes and presents the information.

Thompson, Wilbur. "Internal and External Factors in the Development of Urban Economics." In *Issues in Urban Economics*, edited by H.S. Perloff and L. Wingo, Jr. Baltimore: Johns Hopkins Press, 1968.

After reviewing export base theory and urban-regional income analysis, Thompson discusses possibilities of a theory of urban-regional growth, with his filtering down theory of industrial location.

_____. *A Preface to Urban Economics*. Baltimore: Johns Hopkins Press, 1965.

Tiebout, Charles M. *The Community Economic Base Study*. New York: Committee for Economic Development. Supplementary paper No. 16, 1962.

This background work gives a general view of economic base analysis, which is designed to determine the structure of the local economy. He discusses direct and indirect measures and various multipliers as related to forecasting. He feels that base studies are useful because they focus on volatile sectors separately, and insure consistency among sectors.

_____. "Exports and Regional Economic Growth." "Reply by D.C. North," and "Rejoinder" by Tiebout, *Journal of Political Economy* 64 (April 1956).

This is an argument against Douglas C. North's "Location Theory and Regional Economic Growth," in the same journal, June, 1955. Tiebout says that the necessary condition for economic growth is the development of an export base, but that the larger the region, the greater the simplification of making exports *the* exogenous variable in a regional study.

_____. "Regional and Interregional Input-Output Models: An Appraisal." *Southern Economic Journal* 24 (October 1957): 140-47.

After giving the theoretical framework for a Leontief input-output model, Tiebout discusses the use of I-O modes in local impact studies, regional balance-of-payments studies, and in interregional flow studies. He considers the limitations of I-O analysis, and concludes that most operational regional studies have only tentative results stemming from lack of adequate data and the necessity to make assumptions which vary from reality.

Tinley, John H. and Moglewer, Sidney. "Aerospace Systems Approach Applied to Regional Transportation Planning." Douglas Aircraft Co., Inc., 1965.

Tinley takes the Department of Defense systems approach and suggests its use for transportation planning since it includes the *consequences* of making a specific choice. (Evaluation is based on cost-effectiveness considerations). He mentions the problem of goal determination.

U.S. Department of Commerce, Bureau of Public Roads, Washington, D.C., *Highways and Economic and Social Changes*. Sidney Goldstein and Floyd Thiel, 1964: Government Printing Office.

This work analyzes the results of more than 100 economic impact studies (all done prior to 1960).

U.S. Department of Transportation. Federal Highway Administration and Bureau of Public Roads. "Policy and Procedure Memorandum 20-8." Washington, D.C.: Government Printing Office, 1969.

Vickrey, William S. "Pricing in Urban and Suburban Transportation." *Papers and Proceedings of the American Institute of Planners*. (May 1963).

According to Vickrey, the relevant criterion for pricing is not the function performed, but the degree of congestion that would occur in the absence of pricing. He advocates differential pricing for peak load problems.

Von Boventer, Edwin. "Toward a Theory of Spatial Economic Structure." *Regional Science Association Papers* 10 (1963) 163-87.

This is a review of the literature of location theory.

Voorhees, Alan M. and Associates, Inc. Connecticut Interregional Planning Program. *Analysis of Survey of Personal Attitudes*. Hartford, Conn.: State of Connecticut, 1968.

This study deals with Connecticut residents' attitudes toward housing, residential and job mobility, features most liked, most disliked, and major problems of their town and of the state. It also treats attitudes toward leisure time and recreation. Planning implications, and variations in attitude by income group, family characteristics, and town are considered.

____. *Attitudes of Connecticut People*. Hartford: Connecticut Interregional Planning Program, 1966.

Wachs, Martin and Schofer, Joseph L. "Abstract Values and Concrete Highways." *Traffic Quarterly* 23 (January 1969): 133-56.

A description of program budgeting and its relationship to highway planning. They note that goals and objectives are required before any plan is implemented.

Washington State University, College of Engineering. *Social, Economic, and Environmental Impact of Highway Transportation Facilities on Urban Communities*. Pullman, Wash.: Washington State University, 1968.

Development of a subjective rating schema for evaluating alternatives by use of weights. It considers the overall impact of a transport system and also projects these impacts and their appropriate weights into the future.

____. *A Study of the Social, Economic and Environmental Impact of Highway Transportation Facilities on Urban Communities*. Highway Research Section, Engineering Research Division. Pullman, Washington: Washington State University, 1968.

Wetzel, James R. and Holland, Susan S. "Poverty Areas of our Major Cities." *Monthly Labor Review* (October 1966).

Winfrey, Robley. "Concepts and Application of Engineering Economy in the Highway Field." Highway Research Board, National Academy of Sciences, Washington, D.C., Special Report 56, 1960.

A discussion of the practical aspects of evaluating user cost and benefits. He notes that nonuser benefits should not be included in the body of the engineering economy report.

Witheford, David K. "Highway Impacts on Downtown and Suburban Shopping." Highway Research Board, National Academy of Sciences, Washington, D.C., *Highway Research Record*, 1967, 15-20.

This article compares characteristics of market areas within a fixed travel time from a hypothetical central business district and shopping center. Assuming transportation improvements occur so that larger market areas may be reached in the same time, Witheford finds that the advantage of the shopping center is enhanced.

Yevisaker, Paul N. "The Resident Looks at Community Values." Commissioner, New Jersey Department of Community Affairs, Conference on Transportation and Community Affairs.

Young, Robert C. "Goals and Goal-Setting." *Journal of the American Institute of Planners* 32 (March 1966): 76-85.

A general discussion of the use of goals, objectives, and standards in the planning process. He outlines the process of goal setting and analyzes the conflicts which can arise over them.

Index

UNIVERSITY LIBRARY
NOTTINGHAM

About the Authors

Paul Weiner is Professor of Economics at the University of Connecticut, where he teaches courses in transportation and industrial organization. He received a BA degree at Northeastern University and the MA and PhD at Clark University. Dr. Weiner is serving as a member of the Committee on Economics and the Environment, Highway Research Board, National Academy of Sciences, and is also a Group Council member.

Edward J. Deak has earned the BA and MA at the University of Connecticut. He is currently Assistant Professor of Economics at Fairfield University. Previous publications have appeared in the *Highway Research Record* and the *Rhode Island Business Quarterly*. His field of specialization concerns the technical application of microeconomic theory.